Semilinear Evolution Equations
and Their Applications

Toka Diagana

Semilinear Evolution Equations and Their Applications

 Springer

Toka Diagana
Department of Mathematical Sciences
University of Alabama in Huntsville
Huntsville, AL, USA

Department of Mathematics
Howard University
Washington, DC, USA

King Fahd University of Petroleum and Minerals
Dhahran, Saudi Arabia

ISBN 978-3-030-13114-2   ISBN 978-3-030-00449-1   (eBook)
https://doi.org/10.1007/978-3-030-00449-1

This Springer imprint is published by the registered company Springer Nature Switzerland AG
The registered company address is: Gewerbestrasse 11, 6330 Cham, Switzerland

In memory of Dr. Moussa Diagana

Foreword

Many phenomena in reality are scientifically modeled as evolution equations, and so solutions of those evolution equations are functions in time that are either continuous or discrete, and characterizing quantitative and qualitative properties of the corresponding phenomenon. On the other hand, the study of evolution equations is usually a difficult task, which has led in the past to the development of new branches of mathematics like functional analysis, mathematical analysis, differential equations, stochastic processes, and difference equations. All these achievements are collected nowadays in well-known textbooks, so anyone interested in this knowledge may find it and learn. It is clear that this procedure is still in progress, since new phenomena and problems appear during the current scientific research that lead to new types of evolution equations. Hence recently either new mathematical methods were developed for solving evolution equations or new applications of the known methods have been applied. These achievements are mostly scattered in scientific articles. Consequently, there is a need to collect them in new textbooks and monographs on evolution equations. The monograph, *Semilinear Evolution Equations and Their Applications* by Professor Toka Diagana, fits perfectly in this frame, since on one side, it presents essential known but basic results like from functional analysis, operator theory, semigroups of linear operators for understanding the theory of evolution equations, but on the other side, it serves new results such as singular difference equations, almost periodic functions, fractional evolution equations with many interesting applications for instance arising in population dynamics. This book is a nice continuation of the author's previous one, *Almost Automorphic Type and Almost Periodic Type Functions in Banach Spaces*. The results are well written and the proofs are clearly described. Certainly, it is a pleasure to learn from this book and I recommend it to anyone interested in new trends on evolution and difference equations with applications to partial differential equations and practical problems.

Bratislava, Slovakia Michal Fečkan
March 2018

Preface

This book, which is a continuation of *Almost Automorphic Type and Almost Periodic Type Functions in Banach Spaces*, presents the recent trends and developments upon fractional, first-, and second-order semilinear difference and differential equations on abstract spaces including singular ones. It systematically discusses various stability, existence, and uniqueness results for these semilinear difference and evolution equations. In order to establish the existence results of the book, one makes extensive use of a wide range of tools from nonlinear functional analysis, fractional calculus, and operator theory. Various applications to partial differential equations and population dynamics will be amply discussed.

For the sake of completeness and clarity, ample background materials on Banach and Hilbert spaces including Lebesgue, Sobolev, and Hölder spaces, fixed-point theorems, operator theory, semigroups of operators with special emphasis on analytic semigroups, almost periodic functions, almost periodic sequences on $\mathbb{Z}$ are provided.

Chapter 1 is devoted to the basic material on metric, Banach, and Hilbert spaces needed in the sequel. By definition, every Hilbert space is a Banach—with the converse being untrue. Banach and Hilbert spaces play a central role in many areas and subareas of mathematical analysis as most of the spaces encountered in practical problems turn out to be either Hilbert spaces or Banach spaces. Recall that Banach spaces were named after the Polish mathematician Stefan Banach, who introduced and studied these spaces. Basically, a Banach space is a normed vector space which is complete, that is, every Cauchy sequence in it must converge in it. For instance $\mathbb{Q}$, the field of rational numbers, equipped with the standard absolute value is not complete; its completion is in fact $\mathbb{R}$, the field of real numbers. Standard examples of Banach spaces include, but are not limited to, $\mathbb{R}^d$, $\mathbb{C}^d$, $\ell^p(\mathbb{N})$, L^p, $W^{k,p}$ (Sobolev spaces), $C^{k,\alpha}$ (Holder spaces), $B^s_{p,q}$ (Besov spaces), $H^p(S^1)$ (Hardy spaces), and BMO (functions of bounded mean oscillation).

Chapter 2 reviews basic properties of bounded and unbounded operators as well as their spectral theory. The main references for this chapter are Diagana [47], Bezandry and Diagana [27], and Lunardi [87, 88]. For additional readings on the topics covered in this chapter, we refer to Benzoni [25], Brézis [31, 32],

Conway [38, 39], Gohberg et al. [60], Eidelman et al. [53], Kato [73], Rudin [102], Weidmann [111], and Yosida [112].

In Chap. 3, we collect fundamental results on strongly continuous semigroups and evolution families that are needed in the entire book. The main references for this chapter are the following books: Bezandry and Diagana [27], Chicone and Latushkin [37], Diagana [47], Engel and Nagel [55], Lunardi [87], and Pazy [100], as well as some papers such as Acquistapace and Terreni [8], Baroun et al. [20], and Schnaubelt [105].

The theory of almost periodic functions, which was introduced in the mathematical literature by the Danish mathematician H. Bohr, is an interesting theory that has many applications in various fields including astronomy, signal processing, stochastic differential equations, and harmonic analysis. In Chap. 4, we study the basic properties of almost periodic functions as well as their discrete counterparts needed in the sequel. The main references for this chapter are the following books: Amerio and Prouse [13], Besicovitch [26], Bezandry and Diagana [27], Bohr [29], Corduneanu [40, 41], Diagana [47], Fink [58], Halanay and Rasvan [63], Levitan and Zhikov [82], Levitan [81], N'Guérékata [94, 95], Pankov [99], Zhang [114], Hino et al. [70], as well as papers such as Diagana et al. [49], Diagana [45], Diagana and Pennequin [48], and Fan [56].

The mathematical problem, which consists of studying the existence of solutions to difference equations with almost periodic coefficients, is an important one as almost periodicity, according to Henson et al. [66], is more likely to accurately describe many phenomena occurring in population dynamics than periodicity. In particular, in Diagana et al. [49], the effect of almost periodicity upon population dynamics including the well-known Beverton-Holt model which arises in fishery was investigated. Among others, Diagana et al. made extensive use of dichotomy techniques to obtain sufficient conditions that do guarantee the existence of a globally attracting (Bohr) almost periodic solution to a semilinear system of difference equations on $\mathbb{Z}_+$. These existence results were, next, utilized to study discretely reproducing populations with and without overlapping generations. The main objective in Chap. 5 is twofold. We first extend the above-mentioned theory of almost periodic sequences built by Diagana et al. [49] to $\mathbb{Z}$. Next, we make extensive use of dichotomy techniques to find sufficient conditions for the existence of almost periodic solutions to the semilinear systems of difference equations,

$$x(t+1) = A(t)x(t) + f(t, x(t)), \quad t \in \mathbb{Z}$$

where $A(t)$ is a $k \times k$ almost periodic square matrix, and the function $f : \mathbb{Z} \times \mathbb{R}^k \to \mathbb{R}^k$ is almost periodic in the first variable uniformly in the second one. As in the case of $\mathbb{Z}_+$, our existence results are, subsequently, applied to discretely reproducing populations with overlapping generations.

In Chap. 6, we study and establish the existence of Bohr (respectively, Besicovitch) almost periodic solutions to the following class of singular systems of difference equations,

$$Ax(t + 1) + Bx(t) = f(t, x(t)) \tag{1}$$

where $f : \mathbb{Z} \times \mathbb{R}^N \to \mathbb{R}^N$ is Bohr (respectively, Besicovitch) almost periodic in $t \in \mathbb{Z}$ uniformly in the second variable, and A, B are $N \times N$ square matrices satisfying $\det A = \det B = 0$. Recall that singular difference equations of the form Eq. (1) arise in many applications including optimal control, population dynamics, economics, and numerical analysis [52]. The main result discussed in this chapter can be summarized as follows: if $\lambda A + B$ is invertible for all $\lambda \in S^1 = \{z \in \mathbb{C} : |z| = 1\}$ and if f is Bohr (respectively, Besicovitch) almost periodic in $t \in \mathbb{Z}$ uniformly in the second variable and under some additional conditions, then Eq. (1) has a unique Bohr (respectively, Besicovitch) almost periodic solution. Chapter 6 is organized as follows: Sect. 6.1 serves as an introduction but also provides preliminary tools needed in the sequel. In Sect. 6.2, some preliminary results corresponding to the case $f(t, x(t)) = C(t)$ are obtained. Section 6.3 is devoted to the main results of this chapter. In Sect. 6.4, we make use of the main results in Sect. 6.3 to study the existence of Bohr (respectively, Besicovitch) almost periodic solutions for some second-order (and higher-order) systems of singular difference equations.

Fractional calculus is a generalization of both classical differentiation and integration of non-integer order. Fractional calculus is as old as differential calculus. Fractional differential and integral equations have recently been applied to various areas including engineering, science, finance, applied mathematics, bioengineering, radiative transfer, neutron transport, and the kinetic theory of gases, see, e.g., [16, 33, 35, 36, 71, 74]. Some recent progress in the study of ordinary and partial fractional differential equations has been made, see for instance the following books [4, 6, 19, 51, 77, 91, 101] and [103]. Further, some recent results upon the existence and attractiveness of solutions to various integral equations of two variables have been obtained by many people. In this Chap. 7, we study the existence, uniqueness, estimates, and global asymptotic stability for some classes of fractional integro-differential equations with finite delay. In order to achieve our goal, we make extensive use of some fixed-point theorems as well as the so-called Pachpatte techniques.

In Chap. 8, we study the existence of classical and mild solutions for the Cauchy problem in the inhomogeneous and semilinear cases. The last part of Chap. 8 is devoted to the existence of almost periodic mild solutions to first-order differential equations including nonautonomous ones.

Let $\alpha \in (0, 1]$. Chapter 9 constitutes a brief introduction to the fast-growing theory of fractional evolution equations. More precisely, we study sufficient conditions for the existence of classical (respectively, mild) solutions for the inhomogeneous fractional Cauchy problem

$$\begin{cases} \mathbb{D}_t^\alpha u(t) = Au(t) + f(t), & t > 0 \\ u(0) = u_0 \in \mathbb{X} \end{cases}$$

and its corresponding semilinear evolution equation

$$\begin{cases} \mathbb{D}_t^\alpha u(t) + Au(t) = F(t, u(t)), & t > 0, \\ u(0) = u_0 \in \mathbb{X}, \end{cases}$$

where $\mathbb{D}_t^\alpha$ is the fractional derivative of order α in the sense of Caputo, $A :$ $D(A) \subset \mathbb{X} \mapsto \mathbb{X}$ is a closed linear operator on a complex Banach space $\mathbb{X}$, and $f : [0, \infty) \mapsto \mathbb{X}$ and $F : [0, \infty) \times \mathbb{X} \mapsto \mathbb{X}$ are continuous functions satisfying some additional conditions. Under some appropriate assumptions, various existence results are obtained. The main tools utilized to establish the existence of classical (respectively, mild) solutions to the above-mentioned fractional evolutions are the so-called $(\alpha, \alpha)^\beta$-resolvent families S_α^β and almost sectorial operators $A \in \Sigma_\omega^\gamma(\mathbb{X})$ where $\gamma \in (-1, 0)$ and $0 < \omega < \frac{\pi}{2}$. For more on these tools and related issues, we refer the reader to Keyantuo et al. [75] and Wang et al. [110].

Chapter 10 discusses some existence results for some second-order semilinear evolution equations.

Thermoelastic plate systems play an important role in many applications. For this reason, they have been, in recent years, of great interest to many people. Among other things, the study of the controllability and stability of thermoelastic plate systems has been considered by many people including [14, 18, 24, 44, 65, 79], and [92]. Let $\Omega \subset \mathbb{R}^N$ ($N \geq 1$) be a bounded subset, which is sufficiently regular, and let $a, b : \mathbb{R} \mapsto \mathbb{R}$ be positive functions. In Sect. 10.2, we make extensive use of a wide range of tools including evolution families, real interpolation spaces, and some fixed-point theorems, to study and obtain the existence of almost periodic mild solutions to the following nonautonomous thermoelastic plate systems,

$$\begin{cases} \dfrac{\partial^2 u}{\partial t^2} + \Delta^2 u + a(t)\Delta\theta = f_1(t, \nabla u, \nabla\theta), & \text{if } t \in \mathbb{R}, \ x \in \Omega \\[2mm] \dfrac{\partial\theta}{\partial t} - b(t)\Delta\theta - a(t)\Delta\dfrac{\partial u}{\partial t} = f_2(t, \nabla u, \nabla\theta), & \text{if } t \in \mathbb{R}, \ x \in \Omega, \\[2mm] \theta = u = \Delta u = 0, & \text{if } t \in \mathbb{R}, \ x \in \partial\Omega, \end{cases}$$

where u, θ are the vertical deflection and the variation of temperature of the plate, f_1, f_2 are jointly continuous and (globally) Lipschitz functions. Assuming that the coefficients a, b, f_1, f_2 are almost periodic in the first variable (in $t \in \mathbb{R}$) uniformly in the other ones, it will be shown that the system above has an almost periodic mild solution.

Let $I \subset \mathbb{R}$ be an interval. The main focus in Sect. 10.3 consists of studying the existence of asymptotically almost periodic solutions to some classes of second-order partial functional-differential equations of the form

$$\frac{d}{dt}\left[\varphi'(t) - g(t, \varphi_t)\right] = A\varphi(t) + f(t, \varphi_t), \qquad t \in I,$$

$$\varphi_0 = \psi \in \mathcal{B},$$

$$\varphi'(0) = \xi \in \mathbb{X},$$

where A is the infinitesimal generator of a strongly continuous cosine family $(C(t))_{t \in \mathbb{R}}$ of bounded linear operators on $\mathbb{X}$, the history $\varphi_t : (-\infty, 0] \to \mathbb{X}$, $\varphi_t(\theta) := \varphi(t + \theta)$, belongs to an abstract phase space $\mathscr{B}$ defined axiomatically, and f, g are some appropriate functions. The abstract existence results are subsequently utilized to study the existence of asymptotically almost periodic solutions to some integro-differential equations, which arise in the theory of heat conduction within fading memory materials.

The prerequisite for the book is the Introduction to Real Analysis course material. Although it is primarily intended for beginning graduate students, postgraduates, and researchers, it may also be of interest to nonmathematicians such as physicists and theoretically oriented engineers for instance. Further, it can be used as a graduate text on topics such as operator theory, theory of semigroups, evolution equations, and difference equations and their applications to partial differential equations and practical problems arising in population dynamics.

Toka Diagana

Acknowledgments

First and foremost, I would like to thank all the people and the institutions who, one way or another, contributed in making the writing of this book possible.

I would like to acknowledge the strong support of the King Fahd University of Petroleum and Minerals (KFUPM) during the preparation of this book. I also owe a huge debt of gratitude to the Deanship of Scientific Research for providing an excellent academic environment that is propitious for research activities. Further, I would like to extend my heartfelt thanks to all my colleagues from the mathematics department at KFUPM, especially Prof. Husain S. Al-Attas, the Chair; Prof. Salim Messaoudi; Prof. Amin Boumenir; and Prof. Khaled Furati as well as staff members Anas and Muzaffar for their help and support.

I would like to extend my sincere gratitude and appreciation to Prof. Michal Fečkan for proofreading the book and writing the Foreword. I also extend my thanks to Jamilu Hashim Hassan for careful reading of the manuscript and insightful comments.

I am thankful to all the anonymous reviewers for their useful comments, which greatly enabled me to improve the quality of the book.

I extend my gratitude to Elizabeth Loew and Dahlia Fisch from Springer for their editorial assistance and insightful comments.

Last but certainly not least, I am incredibly grateful to my wife Sally and our wonderful children for their continued support and encouragements.

Contents

1	**Banach and Hilbert Spaces**		1
	1.1	Introduction	1
	1.2	Metric Spaces	2
		1.2.1 Basic Definitions and Examples	2
		1.2.2 Complete Metric Spaces	5
		1.2.3 Continuous Functions	6
		1.2.4 Compact Metric Spaces	7
		1.2.5 Banach Fixed-Point Principle	8
		1.2.6 Equilibrium Points for the Discrete Logistic Equation	10
	1.3	Banach Spaces	10
		1.3.1 Basic Definitions	10
		1.3.2 The Quotient Space	12
		1.3.3 $L^p(\Omega, \mu)$ Spaces	13
		1.3.4 Sobolev Spaces	15
		1.3.5 Embedding Theorems for Sobolev Spaces	17
		1.3.6 Bounded Continuous Functions	19
		1.3.7 Hölder Spaces $C^{k,\alpha}(\overline{\Omega})$	19
		1.3.8 Embedding Theorems for Hölder Spaces	21
		1.3.9 The Dual Space	22
		1.3.10 The Schauder Fixed-Point Theorem	23
		1.3.11 Leray-Schauder Alternative	24
	1.4	Hilbert Spaces	24
		1.4.1 Basic Definitions	24
		1.4.2 Examples of Hilbert Spaces	27
	1.5	Exercises	27
	1.6	Comments	28
2	**Operator Theory**		29
	2.1	Bounded Linear Operators	29
		2.1.1 Basic Definitions	29
		2.1.2 Some Classes of Bounded Operators	31

2.2   Unbounded Linear Operators 33
        2.2.1   Basic Definitions 33
        2.2.2   Examples of Unbounded Operators........................ 34
        2.2.3   The Adjoint Operator in Banach Spaces 35
        2.2.4   Spectral Theory 36
        2.2.5   Sectorial Operators 40
        2.2.6   Examples of Sectorial Operators........................ 40
2.3   Exercises 42
2.4   Comments 43

3 **Semi-Group of Linear Operators** 45
3.1   Introduction 45
3.2   Semi-Group of Operators 45
        3.2.1   Basic Definitions 45
        3.2.2   Analytic Semi-Groups 48
        3.2.3   Hyperbolic Semi-Groups 50
3.3   Evolution Families 51
        3.3.1   Basic Definitions 51
        3.3.2   Acquistapace–Terreni Conditions 52
3.4   Exercises 54
3.5   Comments 56

4 **Almost Periodic Functions and Sequences** 57
4.1   Introduction 57
4.2   Almost Periodic Functions 58
        4.2.1   Basic Definitions 58
        4.2.2   Fourier Series Representation 62
        4.2.3   Composition of Almost Periodic Functions 64
4.3   Asymptotically Almost Periodic Functions 65
        4.3.1   Basic Definitions and Properties 65
        4.3.2   Composition of Asymptotically Almost Periodic
                 Functions 66
4.4   Almost Periodic Sequences 67
        4.4.1   Basic Definitions 67
        4.4.2   Properties of Almost Periodic Sequences 68
        4.4.3   Composition of Almost Periodic Sequences 71
4.5   Asymptotically Almost Periodic Sequences 71
        4.5.1   Basic Definitions 71
        4.5.2   Properties of Asymptotically Almost Periodic Sequences   72
4.6   Exercises 72
4.7   Comments 74

5 **Nonautonomous Difference Equations** 75
5.1   Introduction 75
5.2   Discrete Exponential Dichotomy 77
5.3   The Beverton-Holt Model with Overlapping Generations 81

5.4 Exercises 83
5.5 Comments 83

6 Singular Difference Equations 85
6.1 Introduction 85
6.2 The Case of a Linear Equation 86
    6.2.1 Existence of a Bohr Almost Periodic Solution 87
    6.2.2 Existence of Besicovitch Almost Periodic Solution 89
6.3 The Semilinear Equation 91
6.4 Second-Order Singular Difference Equations 92
6.5 Exercises 94
6.6 Comments 95

7 Fractional Integro-Differential Equations 97
7.1 Introduction 97
7.2 Preliminaries and Notations 98
7.3 Main Results 101
    7.3.1 Existence and Uniqueness 102
    7.3.2 Estimates for Solutions 105
    7.3.3 Global Asymptotic Stability of Solutions 107
7.4 Example 109
7.5 Comments 111

8 First-Order Semilinear Evolution Equations 113
8.1 First-Order Autonomous Evolution Equations 113
    8.1.1 Existence of Mild and Classical Solutions 113
    8.1.2 Existence of Almost Periodic Solutions 115
8.2 Semilinear First-Order Evolution Equations 119
    8.2.1 Existence of Mild and Classical Solutions 119
    8.2.2 Existence Results on the Real Number Line 120
8.3 Nonautonomous First-Order Evolution Equations 121
    8.3.1 Existence of Almost Periodic Mild Solutions 121
8.4 Exercises 124
8.5 Comments 124

9 Semilinear Fractional Evolution Equations 125
9.1 Introduction 125
9.2 Fractional Calculus 126
9.3 Inhomogeneous Fractional Differential Equations 128
    9.3.1 Introduction 128
    9.3.2 Basic Definitions 128
    9.3.3 Existence of Classical and Mild Solutions 131
9.4 Semilinear Fractional Differential Equations 132
    9.4.1 Preliminaries and Notations 132
    9.4.2 Existence Results 133
9.5 Exercises 136
9.6 Comments 137

10   Second-Order Semilinear Evolution Equations 139
   10.1   Introduction 139
   10.2   Almost Periodic Solutions to Some Thermoelastic
           Plate Systems 140
           10.2.1 Introduction 140
           10.2.2 Assumptions on the Coefficients of the
                    Thermoelastic System 142
           10.2.3 Existence of Almost Periodic Solutions 142
   10.3   Existence Results for Some Second-Order Partial
           Functional Differential Equations 151
           10.3.1 Introduction 151
           10.3.2 Preliminaries and Notations 152
           10.3.3 Existence of Local and Global Mild Solutions 154
           10.3.4 Existence of Solutions in Bounded Intervals 155
           10.3.5 Existence of Global Solutions 158
           10.3.6 Existence of Asymptotically Almost Periodic
                    Solutions 161
           10.3.7 Asymptotically Almost Periodic Solutions to Some
                    Second-Order Integro-differential Systems 173
   10.4   Exercises 175
   10.5   Comments 176

Appendix List of Abbreviations and Notations 177

References 181

Index 187

Chapter 1
Banach and Hilbert Spaces

1.1 Introduction

In this chapter we present the basic material on metric, Banach, and Hilbert spaces needed in the sequel. By design, every Hilbert space is a Banach space—with the converse being untrue. Banach and Hilbert spaces play a central role in many areas and subareas of mathematical analysis as most of the spaces encountered and utilized in practical problems turn out to be either Hilbert spaces or Banach spaces.

Recall that Banach spaces were named after the Polish mathematician Stefan Banach who introduced them in the mathematical literature around 1920–1922. Basically, a Banach space is a normed vector space that is complete, that is, every Cauchy sequence in it must converge in it. For instance $\mathbb{Q}$ the field of rational numbers equipped with the standard absolute value is not complete; its completion is in fact $\mathbb{R}$, the field of real numbers. Standard examples of Banach spaces include, but are not limited to, $\mathbb{R}^d$, $\mathbb{C}^d$, $\ell^p(\mathbb{N})$, L^p (Lebesgue spaces), $W^{k,p}$ (Sobolev spaces), $C^{k,\alpha}$ (Hölder spaces), $B^s_{p,q}$ (Besov spaces), $H^p(\mathbb{S}^1)$ (Hardy spaces), and BMO (functions of bounded mean oscillation)—when they are equipped with their respective standard norms.

In this chapter, we first study some of the basic properties of metric spaces and then use these to deduce those of Banach and Hilbert spaces.

One should stress upon the fact that the introductory material presented in this chapter can be found in any good book in (nonlinear) functional analysis. Consequently, some proofs will be omitted.

© Springer Nature Switzerland AG 2018
T. Diagana, *Semilinear Evolution Equations and Their Applications*,
https://doi.org/10.1007/978-3-030-00449-1_1

1.2   Metric Spaces

This section is essentially devoted to metric spaces and their basic properties. Among other things, the following notions will be introduced and studied in the context of metric spaces: convergence, completeness, continuity, compact metric spaces, and the Banach fixed-point principle.

1.2.1   Basic Definitions and Examples

Definition 1.1 A pair (M, d) consisting of a nonempty set M and a mapping (metric or distance) $d : M \times M \mapsto [0, \infty)$ is called a metric space, if the mapping d fulfills the following properties,

i) $d(x, y) = 0$ if and only if $x = y$;
ii) $d(x, y) = d(y, x)$; and
iii) $d(x, z) \leq d(x, y) + d(y, z)$

for all $x, y, z \in M$.

The inequality iii) appearing in Definition 1.1 is commonly known as the *triangle inequality*. Further, elements of the set M are called points and the quantity $d(x, y)$ is referred to as the distance between the points $x, y \in M$.

If (M, d) is a metric space, then using the triangle inequality, it can be easily shown that the following property holds,

$$\left| d(x, y) - d(y, z) \right| \leq d(x, z)$$

for all $x, y, z \in M$.

Example 1.2

(1) Let d_0 and d_1 be two metrics upon a nonempty set M. Consider the mapping d defined by, $d(x, y) = \alpha d_0(x, y) + \beta d_1(x, y)$ for all $x, y \in M$, where $\alpha, \beta \geq 0$ and $\alpha + \beta = 1$. It is not hard to see that the pair (M, d) is a metric space.
(2) Let M be an arbitrary nonempty set which we endow with the so-called discrete metric d_s defined by

$$d_s(x, y) := \begin{cases} 0 & \text{if } x = y \\ \\ 1 & \text{if } x \neq y \end{cases}$$

It can be easily shown that (M, d_s) is a metric space.

(3) Fix $p \geq 1$. Then $(\mathbb{R}^n, d_p)$ is a metric space, where

$$d_p(\alpha, \beta) := \left(\sum_{k=1}^{n} |\alpha_k - \beta_k|^p \right)^{1/p}$$

for all $\alpha = (\alpha_1, \alpha_2, ..., \alpha_n)$, $\beta = (\beta_1, \beta_2, ..., \beta_n) \in \mathbb{R}^n$.

(4) Consider the unit circle, $\mathbb{S}^1 = \{z \in \mathbb{C} : |z| = 1\}$ and let $\mathscr{A}(\mathbb{S}^1)$ stand for the collection of all functions $f : \mathbb{S}^1 \mapsto \mathbb{C}$ for which

$$\int_0^{2\pi} |f(e^{i\theta})|^2 d\theta < \infty.$$

It can be easily seen that $(\mathscr{A}(\mathbb{S}^1), \rho)$ is a metric space, where the metric ρ is defined by,

$$\rho(f, g) := \left(\int_0^{2\pi} |f(e^{i\theta}) - g(e^{i\theta})|^2 d\theta \right)^{\frac{1}{2}}$$

for all $f, g \in \mathscr{A}(\mathbb{S}^1)$.

As usual, if (M, d) is a metric space, then the metric d enables us to define the notions of balls and spheres in M. Indeed, the (open) ball centered at $x \in M$ with radius $r > 0$ is defined by $B(x, r) = \{y \in M : d(x, y) < r\}$. Similarly, the (closed) ball centered at $x \in M$ with radius $r \geq 0$ is defined by $\overline{B}(x, r) = \{y \in M : d(x, y) \leq r\}$. By the sphere $S(x, r)$, centered at $x \in M$ with radius $r \geq 0$, we mean the set of all points defined by $S(x, r) = \{y \in M : d(x, y) = r\}$.

Definition 1.3 Let (M, d) be a metric space and let $\mathscr{O} \subset M$ be a subset. The set $\mathscr{O}$ is said to be an open set if for all $x \in M$, there exists $r > 0$ such that $B(x, r) \subset \mathscr{O}$.

Classical examples of open sets in a metric space (M, d) include M itself and the empty set, $\emptyset$. Recall that arbitrary unions of open sets of M are also open sets of M. Further, finite intersections of open sets of M are also open sets of M.

Definition 1.4 Let (M, d) be a metric space and let $\mathscr{O} \subset M$ be a subset. A point $x \in M$ is said to be an interior point of $\mathscr{O}$ if and only if there exists $r > 0$ such that $B(x, r) \subset \mathscr{O}$. The collection of all interior points of $\mathscr{O}$ is denoted $\text{Int}(\mathscr{O})$.

It can be shown that a subset $\mathscr{O}$ of M is open if and only if it contains all of its interior points, that is, $\mathscr{O} = \text{Int}(\mathscr{O})$.

Definition 1.5 Let (M, d) be a metric space and let $\mathscr{O} \subset M$ be a subset. The set $\mathscr{O}$ is said to be a closed set if its complement $\mathscr{O}^C = M \setminus \mathscr{O}$ is an open set.

Classical examples of closed sets of a metric space (M, d) include singletons $\{x\}$, M, and $\emptyset$.

Definition 1.6 Let (M, d) be a metric space and let $\mathcal{O} \subset M$ be a subset. A point $x \in M$ is said to be an adherent point of $\mathcal{O}$ if and only if for all $r > 0$, the following holds, $B(x, r) \cap \mathcal{O} \neq \{\emptyset\}$. The collection of all adherent points of $\mathcal{O}$ is called the closure of $\mathcal{O}$ and is denoted $\overline{\mathcal{O}}$.

It can be shown that a subset $\mathcal{O}$ of M is closed if and only if it contains all of its adherent points, that is, $\mathcal{O} = \overline{\mathcal{O}}$. In other words, $\mathcal{O}$ is closed if and only if for any sequence $(x_n)_{n \in \mathbb{N}} \subset \mathcal{O}$ such that $d(x_n, x) \to 0$ for some $x \in M$ as $n \to \infty$, then one must have $x \in \mathcal{O}$.

Definition 1.7 Let (M, d) be a metric space and let $\mathcal{O} \subset M$ be a subset. The subset $\mathcal{O}$ is said to be bounded it is included in some ball $B(x, r)$. Otherwise, the set $\mathcal{O}$ is said to be unbounded.

It is easy to see that $\mathcal{O} \subset M$ is bounded if and only if its diameter, $\operatorname{diam}(\mathcal{O}) :=$ $\sup_{x, y \in \mathcal{O}} d(x, y)$, is finite, that is, $\operatorname{diam}(\mathcal{O}) < \infty$.

If (M, d) is a metric space, then the metric d enables us to define the notion of convergence in M.

Definition 1.8 Let (M, d) be a metric space. A sequence $(x_n)_{n \in \mathbb{N}} \subset M$ is said to converge to some $x \in M$ with respect to the metric d, if $d(x_n, x) \to 0$ as $n \to \infty$. Equivalently, for every $\varepsilon > 0$, there exists $N \in \mathbb{N}$ such that $d(x_n, x) < \varepsilon$ for all $n \geq N$.

If a sequence $(x_n)_{n \in \mathbb{N}} \subset M$ converges to some $x \in M$ with respect to the metric d, then we write $\lim_{n \to \infty} x_n = x$.

Proposition 1.9 *Let (M, d) be a metric space. If a sequence $(x_n)_{n \in \mathbb{N}} \subset M$ converges, then its limit is unique.*

Proof Suppose $(x_n)_{n \in \mathbb{N}} \subset M$ converges to two limits $x, y \in M$. Then, using the triangle inequality it follows that $0 \leq d(x, y) \leq d(x, x_n) + d(x_n, y)$. Letting $n \to \infty$ in the previous inequality, it follows that $d(x, y) = 0$ which yields $x = y$.

Definition 1.10 Let (M, d) be a metric space. A sequence $(x_n)_{n \in \mathbb{N}} \subset M$ is called a Cauchy sequence, if for every $\varepsilon > 0$, there exists $N \in \mathbb{N}$ such that $d(x_n, x_m) < \varepsilon$ for all $n, m \geq N$.

Proposition 1.11 *Let (M, d) be a metric space. Every convergent sequence is a Cauchy sequence. Further, every Cauchy sequence is bounded.*

Proof

i) Let $(x_n)_{n \in \mathbb{N}}$ be a convergent sequence in the metric space (M, d). This means that there exists $x \in M$ such that $d(x_n, x) \to 0$ as $n \to \infty$. Equivalently, for all $\varepsilon > 0$, there exists $N \in \mathbb{N}$ such that $d(x_n, x) < \frac{\varepsilon}{2}$ for all $n \geq N$. Now, using the triangle inequality it follows that $d(x_n, x_m) \leq d(x_n, x) + d(x, x_m) < \frac{\varepsilon}{2} + \frac{\varepsilon}{2} = \varepsilon$ for all $n, m \geq N$. Consequently, $(x_n)_{n \in \mathbb{N}}$ is a Cauchy sequence.

ii) Suppose that $(x_n)_{n \in \mathbb{N}}$ is a Cauchy sequence. For $\varepsilon = 1$, there exists $N \in \mathbb{N}$ such that $d(x_n, x_m) < 1$ for $n, m \geq N$. In particular, $d(x_n, x_N) < 1$ for all $n \geq N$. Setting $r = 1 + \max\{1, d(x_1, x_N), d(x_2, x_N), ..., d(x_{N-1}, x_N)\}$, one can easily see that $(x_n)_{n \in \mathbb{N}} \subset B(x_N, r)$ which yields the sequence $(x_n)_{n \in \mathbb{N}}$ is bounded.

1.2.2 Complete Metric Spaces

Definition 1.12 A metric space (M, d) is said to be complete, if every Cauchy sequence in it converges in it.

Classical examples of complete metric spaces include $\mathbb{R}^n$ equipped with its corresponding Euclidean metric defined by

$$d(x, y) = \left(\sum_{k=1}^{n} |x_k - y_k|^2 \right)^{\frac{1}{2}}$$

for all $x = (x_1, x_2, ..., x_n)$, $y = (y_1, y_2..., y_n) \in \mathbb{R}^n$, and $BC(\mathbb{R}, M)$ the collection of all bounded continuous functions which go from $\mathbb{R}$ into a complete metric space (M, d), when it is equipped with the sup norm metric, $d_\infty(f, g) = \sup_{t \in \mathbb{R}} d(f(t), g(t))$ for all $f, g \in BC(\mathbb{R}, M)$, etc.

A classical example of a metric space which is not complete is $\mathbb{Q}$; the field of rational numbers; when it is equipped with the standard absolute value defined by $d_0(x, y) = |x - y|$ for all $x, y \in \mathbb{Q}$. There are obviously various ways of constructing a Cauchy sequence in $\mathbb{Q}$ which diverges. Let us exhibit one here. Indeed, consider the recurrent sequence $(x_n)_{n \in \mathbb{N}}$ given by,

$$\begin{cases} x_1 = q \in \mathbb{N}, \\[2mm] x_{n+1} = \dfrac{2}{3}x_n + \dfrac{1}{x_n}, & \text{for all } n \in \mathbb{N}. \end{cases}$$

It is not hard to show that not only $(x_n)_{n \in \mathbb{N}}$ is a sequence of rational numbers but also that $\lim_{n \to \infty} x_n = \sqrt{3} \in \mathbb{R} \setminus \mathbb{Q}$. This shows that $(\mathbb{Q}, d_0)$ is not complete.

The larger question regarding the completion of a given arbitrary metric space (M, d) is the following: if (M, d) is not complete, are there ways of making it complete? The answer is a "yes." Indeed, suppose that (M, d) is not complete and let

$$CS(M, d) := \Big\{ \text{all Cauchy sequences in } (M, d) \Big\}.$$

Define an equivalence relation upon elements of $CS(M, d)$ as follows: two sequences $(x_n)_{n \in \mathbb{N}}$, $(y_n)_{n \in \mathbb{N}}$ in $CS(M, d)$ are said to be equivalent which we denote $(x_n)_{n \in \mathbb{N}} \sim (y_n)_{n \in \mathbb{N}}$ if and only if, for $\varepsilon > 0$, there exists $N \in \mathbb{N}$ such that

$$d(x_n, y_m) < \varepsilon \text{ for all } n, m \geq N.$$

Consider

$$\widetilde{CS(M, d)} = CS(M, d)/ \sim$$

and define the mapping $\tilde{d} : \widetilde{CS(M, d)} \times \widetilde{CS(M, d)} \mapsto [0, \infty)$ as follows: if $x, y \in \widetilde{CS(M, d)}$, that is, $x = [(x_n)_n]$ and $y = [(y_n)_{n \in \mathbb{N}}]$ (equivalence classes of $(x_n)_{n \in \mathbb{N}}$ and $(y_n)_{n \in \mathbb{N}}$), then

$$\tilde{d}(x, y) = \lim_{n \to \infty} d(x_n, y_n).$$

It is easy to check that $\tilde{d}$ is a metric upon $\widetilde{CS(M, d)}$. Moreover, the metric space $(\widetilde{CS(M, d)}, \tilde{d})$, by construction, is complete.

1.2.3 Continuous Functions

In the sequel, the pairs (M, d), (M', ρ), (M_1, d_1), (M_2, d_2), and (M_3, d_3) stand for metric spaces.

Definition 1.13 A function $f : (M, d) \mapsto (M', \rho)$ is said to be continuous at $x_0 \in M$, if for all $\varepsilon > 0$ there exists $\delta > 0$ such that for all $x \in M$, $d(x_0, x) < \delta$ yields $\rho(f(x_0), f(x)) < \varepsilon$. The function f is said to be continuous on M, if it is continuous at each point of M.

Recall that the continuity of a function $f : (M, d) \mapsto (M', \rho)$ at $x_0 \in M$ is equivalent to its sequential continuity at $x_0 \in M$, that is, for any arbitrary sequence $(x_n)_{n \in \mathbb{N}} \subset M$ that converges to some x_0, we have that $f(x_n)$ converges to $f(x_0)$. In general, it is easier to prove the continuity of a function using the sequential continuity than the general definition of continuity given in Definition 1.13.

We have the following composition result for continuous functions on metric spaces whose proof is left to the reader as an exercise.

Proposition 1.14 Let $f : M_1 \mapsto M_2$ and $g : M_2 \mapsto M_3$ be given functions. Let $x_0 \in M_1$ be such that f is continuous at x_0 and that g is continuous at $f(x_0)$. Then, $g \circ f$, the composition of f with g, is also continuous at x_0.

Definition 1.15 A function $f : (M, d) \mapsto (M', \rho)$ is said to be uniformly continuous if, for all $\varepsilon > 0$, there exists $\delta > 0$ such that for all $x, y \in M$, $d(x, y) < \delta$ yields $\rho(f(x), f(y)) < \varepsilon$.

Obviously, every uniformly continuous function is continuous. However, the converse, except in the case when M is a compact metric space, is in general not true (see for instance Theorem 1.20).

1.2.4 Compact Metric Spaces

Definition 1.16 An open covering of (M, d) is a collection of open sets whose union is M.

Definition 1.17 A metric space (M, d) is said to be compact if every open covering of it has a finite subcovering.

Definition 1.18 A subset $\mathcal{O}$ of a metric space (M, d) is called relatively compact if its closure $\overline{\mathcal{O}}$ is a compact subset of M.

Theorem 1.19 *A metric space (M, d) is compact if and only if every sequence of elements of M has a subsequence which converges.*

Proof The proof is left to the reader as an exercise.

Theorem 1.20 *If (M, d) is a compact metric space and if $f : M \mapsto M'$ is a continuous function, then f is uniformly continuous.*

Proof Let $\varepsilon > 0$ and $x \in M$ be given. Using the fact that f is continuous, we deduce that there exists $\delta_x > 0$ such that $d(x, y) < \delta_x$ yields $\rho(f(x), f(y)) < \frac{\varepsilon}{2}$. Since M is compact, it follows that there exists a finite number of points $x_1, x_1, ..., x_d$ such that the following holds,

$$M \subset \bigcup_{j=1}^{d} B\left(x_j, \frac{\delta_{x_j}}{2}\right).$$

Now let $\delta = \frac{1}{2} \min\{\delta_{x_j} : j = 1, 2, ..., d\}$. Clearly, $\delta > 0$. Further, for all $(x, y) \in M \times M$ satisfying $d(x, y) < \delta$, one can find $j_0 \in \{1, 2, ..., d\}$ such that x and y belong to $B(x_{j_0}, \delta_{j_0})$, which yields $\rho(f(x), f(x_{j_0})) < \frac{\varepsilon}{2}$ and $\rho(f(y), f(x_{j_0})) < \frac{\varepsilon}{2}$. Using the triangle inequality it follows that,

$$\rho(f(x), f(y)) \leq \rho(f(x), f(x_{j_0})) + \rho(f(x_{j_0}), f(y)) < \varepsilon,$$

which shows that the function f is uniformly continuous.

Let $C(M, M')$ denote the collection of all continuous functions from M into M'. If $M' = \mathbb{C}$, then $C(M, \mathbb{C})$ will be denoted by $C(M)$. If M is compact, then $C(M)$ is a metric space when it is equipped with the metric, $d_\infty(u, v) = \max_{x \in M} |u(x) - v(x)|$ for all $u, v \in C(M)$.

Definition 1.21 A sub-collection $\Gamma \subset C(M)$ is called uniformly bounded if there exists $C > 0$ such that $|u(x)| \leq C$ for every $x \in M$ and $u \in \Gamma$.

Definition 1.22 A sub-collection $\Gamma \subset C(M, M')$ is said to be equi-continuous if for all $x_0 \in M$ and for all $\varepsilon > 0$ there exists $\delta = \delta(x_0, \varepsilon) > 0$ such that $d(x, x_0) < \delta$ yields $\rho(f(x), f(x_0)) < \varepsilon$ for all $f \in \Gamma$.

Example 1.23 Let $\Gamma \subset C(M)$ be defined as the family of all K-Lipschitz functions on M where $K \geq 0$. That is, all functions $u : M \mapsto \mathbb{C}$ such that $|u(x) - u(y)| \leq Kd(x, y)$ for all $x, y \in M$. It can be easily shown that the family Γ is equi-continuous.

Theorem 1.24 (Arzelà-Ascoli Theorem) *A sub-collection* $\Gamma \subset C(M)$ *is relatively compact if and only if,*

(a) Γ is equi-continuous; and
(b) Γ is uniformly bounded.

Proof The proof is left to the reader as an exercise.

Corollary 1.25 *A sub-collection* $\Gamma \subset C(M)$ *is compact if and only if it is closed, uniformly bounded, and equi-continuous.*

Proof The proof is left to the reader as an exercise.

1.2.5   Banach Fixed-Point Principle

Definition 1.26 A mapping $T : (M, d) \mapsto (M', \rho)$ is said to be Lipschitz if there exists $K \geq 0$ (Lipschitz constant) such that

$$\rho(T(x), T(y)) \leq Kd(x, y)$$

for all $x, y \in M$. In the case when $0 \leq K < 1$, then the map T is said to be a strict contraction.

Example 1.27 Let $\mathscr{O} \subset M$ be a subset. If $x \in M$, ones defines the distance between the point x and the set $\mathscr{O}$ as follows,

$$d(x, \mathscr{O}) := \inf_{y \in \mathscr{O}} d(x, y).$$

Proposition 1.28 *The map* $D : M \mapsto \mathbb{R}_+, x \mapsto D(x) := d(x, \mathcal{O})$ *is Lipschitz with* 1 *as its Lipschitz constant. That is,*

$$|d(x, \mathcal{O}) - d(y, \mathcal{O})| \leq d(x, y)$$

for all $x, y \in M$.

Proof Using the triangle inequality and a property of the infimum it follows that $D(x) \leq d(x, e) \leq d(x, y) + d(y, e)$ for all $x, y \in M$ and $e \in \mathcal{O}$, which yields $D(x) - d(x, y) \leq d(y, e)$ for all $x, y \in M$ and $e \in \mathcal{O}$. Using the fact that the infimum is the greatest element that is less than or equal to all elements, we deduce that $D(x) - d(x, y) \leq D(y)$ which in turn yields $D(x) - D(y) \leq d(x, y)$ for all $x, y \in M$. Since x and y are arbitrary elements of M, replacing x with y in the previous inequality, one gets, $D(y) - D(x) \leq d(y, x) = d(x, y)$ for all $x, y \in M$. Combing the last two inequalities, we deduce that $|D(x) - D(y)| \leq d(x, y)$ for all $x, y \in M$.

If $T : (M, d) \mapsto (M, d)$ is a mapping, then one defines its fixed-points by $F_T = \{x \in M : T(x) = x\}$.

Remark 1.29 Note that if T is a strict contraction, then F_T cannot contain more than one element. Indeed, suppose $x = Tx$ and $y = Ty$, then $d(x, y) = d(T(x), T(y)) \leq Kd(x, y) < d(x, y)$, this impossible. Consequently, F_T cannot contain more than one element.

Theorem 1.30 (Banach Fixed-Point Theorem [47]) *If* (M, d) *is a complete metric space and if* $T : M \mapsto M$ *is a strict contraction, then it has a unique fixed-point.*

Proof Let $x_0 \in M$. Define the sequence $x_n = T^n x_0$ which yields $x_{n+1} = Tx_n$ and $x_n = Tx_{n-1}$. Consequently, $d(x_{n+1}, x_n) \leq K^n d(x_0, x_1)$ where $0 \leq K < 1$ is the Lipschitz constant. Similarly, for all $n, m \in \mathbb{N}$ with $n > m$, we have

$$d(x_n, x_m) \leq d(x_{n+1}, x_n) + d(x_n, x_{n-1}) + \ldots + d(x_m, x_{m-1})$$

$$\leq \left(K^n + K^{n-1} + \ldots + K^{m-1}\right) d(x_1, x_0)$$

$$\leq \frac{K^n}{1 - K} d(x_1, x_0).$$

This yields the sequence $(x_n)_{n \in \mathbb{N}} \subset M$ which is a Cauchy sequence and since (M, d) is complete, it follows that there exists $x \in M$ such that $\lim_{n \to \infty} d(x_n, x) = 0$. Since T is continuous, it follows that $Tx = \lim_{n \to \infty} Tx_n = \lim_{n \to \infty} x_{n+1} = x$ which yields $x \in F_T$. Using Remark 1.29, we deduce that $F_T = \{x\}$.

1.2.6   Equilibrium Points for the Discrete Logistic Equation

This subsection is an application to the Banach fixed-point theorem and is based upon [72, Chapter 3]. Indeed, we make use of the Banach fixed-point theorem to determine equilibrium points of the (discrete) logistic equation. One of the most celebrated discrete dynamical systems arising in dynamic of population is that of the logistic equation, which is given by the following nonlinear difference equation,

$$x(t + 1) = 4\sigma x(t)(1 - x(t)),$$

where $\sigma \in [0, 1]$ and $x(t) \in [0, 1]$ for all $t \in \mathbb{N}$.

In many concrete applications, $x(t)$ stands for the size of the population being studied at the generation t of a reproducing population. Recall that the linear part of the logistic equation, that is, $x(t + 1) = 4\sigma x(t)$, describes, depending upon σ, the exponential growth (if $\sigma > \frac{1}{4}$) or decay (if $\sigma < \frac{1}{4}$) of a population subject to constant birth or death rate.

Clearly, the discrete logistic equation is of the form, $x(t + 1) = Sx(t)$, where the function S (logistic map) is defined by

$$S : [0, 1] \mapsto [0, 1], \quad x \mapsto 4\sigma x(1 - x).$$

Obviously, equilibrium points of the logistic equation correspond to the fixed-points of the logistic map S. It can be easily shown that for $\sigma \in [0, \frac{1}{4}]$, the point $x_0 = 0$ is the only fixed-point (equilibrium point of the discrete logistic equation) of S. For $\sigma \in (\frac{1}{4}, 1]$, fixed-points of the logistic map S are given by the points of the form, $x(\sigma) := 1 - (4\sigma)^{-1}$. Consequently, the logistic map S has infinity many fixed-points (or equilibrium points for the logistic equation) given by

$$F_S = \left\{0\right\} \cup \left\{x_\sigma := 1 - (4\sigma)^{-1} : \sigma \in (\frac{1}{4}, 1]\right\}.$$

1.3   Banach Spaces

In the rest of this book, unless otherwise stated, $\mathbb{F}$ stands for the field of real numbers $\mathbb{R}$ or the field of complex numbers $\mathbb{C}$.

1.3.1   Basic Definitions

Let $\mathbb{X}$ be a vector space over the field $\mathbb{F}$. A norm on $\mathbb{X}$ is a mapping $\| \cdot \| : \mathbb{X} \mapsto \mathbb{R}_+$ satisfying the following properties,

i) $\|x\| = 0$ if and only if $x = 0$;

ii) $\|\lambda x\| = |\lambda| \, \|x\|$; and

iii) $\|x + y\| \leq \|x\| + \|y\|$

for all $x, y \in \mathbb{X}$ and $\lambda \in \mathbb{F}$.

The pair $(\mathbb{X}, \|\cdot\|)$ is then called a normed vector space. From a given normed vector space $(\mathbb{X}, \|\cdot\|)$, one can construct a metric space $(\mathbb{X}, d)$, where the metric d is defined by $d(x, y) := \|x - y\|$ for all $x, y \in \mathbb{X}$.

In what follows, we introduce two important notions which play a crucial role in many fields: separable and uniformly convex normed vector spaces. For that, we need to introduce a few notions.

Definition 1.31 Let $(\mathbb{X}, \|\cdot\|)$ be a normed vector space and let $\mathscr{D} \subset \mathbb{X}$ be a subset. Then, $\mathscr{D}$ is said to be *dense*, if $\mathbb{X} = \overline{\mathscr{D}}$. Equivalently, for each $x \in \mathbb{X}$, there exists a sequence $(x_n)_{n \in \mathbb{N}} \subset \mathscr{D}$ such that $d(x_n, x) = \|x_n - x\| \to 0$ as $n \to \infty$.

Definition 1.32 A set $\mathscr{D}$ is called *countable* if it is finite or has the same cardinality as $\mathbb{N}$ (i.e., there exists a bijection between $\mathscr{D}$ and $\mathbb{N}$). A set $\mathscr{D}$ is called *uncountable* if it is infinite and not countable.

Definition 1.33 A normed vector space $(\mathbb{X}, \|\cdot\|)$ is said to be separable if it contains a countable dense subset $\mathscr{D}$.

Classical examples of separable normed vector spaces include, but are not limited to, $(\mathbb{R}, |\cdot|)$, $(\mathbb{C}, |\cdot|)$, and $(\ell^p(\mathbb{N}), \|\cdot\|_p)$ for $1 \leq p < \infty$, where $\ell^p(\mathbb{N})$ is vector space consisting of all sequences $x = (x_n)_{n \in \mathbb{N}}$ with $x_n \in \mathbb{C}$ for all $n \in \mathbb{N}$ and

$$\|x\|_p := \left(\sum_{k=1}^{\infty} |x_k|^p \right)^{1/p} < \infty.$$

It is also well known that $\ell^\infty(\mathbb{N})$, the vector space of all bounded sequences (the vector space consisting of all sequences $x = (x_n)_{n \in \mathbb{N}}$ with $x_n \in \mathbb{C}$ for all $n \in \mathbb{N}$ such that $|x_n| \leq M$ for all $n \in \mathbb{N}$ with $M \geq 0$ being a constant), is not separable, when it is equipped with its natural sup-norm $\|\cdot\|_\infty$ defined, for each $x = (x_n)_{n \in \mathbb{N}} \in \ell^\infty(\mathbb{N})$, by

$$\|x\|_\infty = \sup_{n \in \mathbb{N}} |x_n|.$$

Definition 1.34 A normed vector space $(\mathbb{X}, \|\cdot\|)$ is said to be uniformly convex if for each $\varepsilon > 0$ there exists $\delta > 0$ such that

$$x, y \in \mathbb{X}, \ \|x\| \leq 1, \ \|y\| \leq 1 \text{ and } \|x - y\| \geq \varepsilon \text{ yields } \left\| \frac{x + y}{2} \right\| < 1 - \delta.$$

While the normed vector spaces $(\mathbb{R}, |\cdot|)$, $(\mathbb{C}, |\cdot|)$, and $(\ell^p(\mathbb{N}), \|\cdot\|_p)$ for $1 < p < \infty$ are uniformly convex, $\ell^\infty(\mathbb{N})$ is not.

Definition 1.35 A normed vector space $(\mathbb{X}, \|\cdot\|)$ is said to be a *Banach space* if the metric space $(\mathbb{X}, d)$, where $d(x, y) := \|x - y\|$ for all $x, y \in \mathbb{X}$, is complete.

Classical examples of Banach spaces include finite-dimensional normed vector spaces. Obviously, there are plenty of normed vector spaces which are not Banach spaces (not complete). Indeed, for $a < b$, consider $C[a, b]$, the collection of all continuous functions $f : [a, b] \mapsto \mathbb{R}$. It is clear that $C[a, b]$ is a vector space over $\mathbb{R}$ which we equip with the norm given, for $p \in [1, \infty)$, by

$$\|f\|_p := \left(\int_a^b |f(t)|^p dt \right)^{\frac{1}{p}} \text{ for all } f \in C[a, b].$$

It is not hard to see that the normed vector space $(C[a, b], \|\cdot\|_p)$ is not complete.

Obviously, Banach spaces are more interesting for applications than incomplete normed vector spaces. Consequently, in what follows, our main focus will be on Banach spaces and their basic properties.

The proof of the next theorem presents no difficulty and hence is left to the reader as an exercise.

Theorem 1.36 *If $(\mathbb{X}, \|\cdot\|)$ is a Banach space and if $\mathbb{Y} \subset \mathbb{X}$ is a subspace, then $(\mathbb{Y}, \|\cdot\|)$ is a Banach space if and only if $\mathbb{Y}$ is closed.*

1.3.2   The Quotient Space

Definition 1.37 Let $\mathbb{L}$ be a subspace of the vector space $\mathbb{X}$. The cosets of $\mathbb{L}$ are defined by the sets, $[x] = x + \mathbb{L} = \{x + \ell : \ell \in \mathbb{L}\}$.

Define the quotient $\mathbb{X} \setminus \mathbb{L}$ as follows $\mathbb{X} \setminus \mathbb{L} = \{[x] : x \in \mathbb{X}\}$. The canonical projection of $\mathbb{X}$ onto $\mathbb{X} \setminus \mathbb{L}$ is defined by $\pi : \mathbb{X} \mapsto \mathbb{X} \setminus \mathbb{L}, x \mapsto [x]$. It is not hard to see that π is surjective and that $Ker(\pi) = \{x \in \mathbb{X} : \pi(x) = 0\} = \mathbb{L}$. Further, $\mathbb{X} \setminus \mathbb{L}$ is a vector space over $\mathbb{F}$ ($[x + y] = [x] + [y]$ and $[\lambda x] = \lambda[x]$ for all $x, y \in \mathbb{X}$ and $\lambda \in \mathbb{F}$) called the quotient space. Furthermore, the mapping $\|\cdot\| : \mathbb{X} \setminus \mathbb{L} \mapsto [0, \infty)$ defined by

$$\|[x]\| = \|x + \mathbb{L}\| = d(x, \mathbb{L}) = \inf_{y \in \mathbb{L}} \|x - y\|$$

for each $[x] \in \mathbb{X} \setminus \mathbb{L}$, is a norm on the quotient vector space $\mathbb{X} \setminus \mathbb{L}$. Indeed, for all $x, y \in \mathbb{X}$ and $\lambda \in \mathbb{F}$, using the fact that $\mathbb{L}$ is closed, we have $d(x, \mathbb{L}) = 0$ if and only if $x \in \mathbb{L}$. Thus $\|[x]\| = \|x + \mathbb{L}\| = 0$ if and only if $[x] = x + \mathbb{L} = 0 + \mathbb{L}$.

Suppose $\lambda \neq 0$. Then, we have

$$\|\lambda[x]\| = \|\lambda(x + \mathbb{L})\| = d(\lambda x, \mathbb{L}) = d(\lambda x, \lambda \mathbb{L}) = |\lambda| d(x, \mathbb{L}) = |\lambda| \|[x]\|.$$

Now, if $\lambda = 0$, it easily follows that

$$\|0(x + \mathbb{L})\| = \|0 + \mathbb{L}\| = 0 = |0| \|x + \mathbb{L}\|.$$

Now let $x_1, y_1 \in \mathbb{L}$, then

$$\begin{aligned}
\|(x + \mathbb{L}) + (y + \mathbb{L})\| &= \|(x + y) + \mathbb{L}\| \\
&\leq \|x + y + x_1 + y_1\| \\
&\leq \|x + x_1\| + \|y + y_1\|,
\end{aligned}$$

which yields

$$\|(x + \mathbb{L}) + (y + \mathbb{L})\| \leq \|x + \mathbb{L}\| + \|y + \mathbb{L}\|$$

That is,

$$\|[x] + [y]\| \leq \|[x]\| + \|[y]\|.$$

Recall that if $\mathbb{L}$ a closed subspace of the normed vector space $(\mathbb{X}, \|\cdot\|)$, then the canonical projection π is linear and continuous as $\|\pi(x)\| \leq \|x\|$ for all $x \in \mathbb{X}$.

Theorem 1.38 *Let $(\mathbb{X}, \|\cdot\|)$ be Banach space and let $\mathbb{L} \subset \mathbb{X}$ is a closed subspace. Then the quotient normed vector space $(\mathbb{X} \setminus \mathbb{L}, \|\cdot\|)$ is a Banach space.*

1.3.3 $L^p(\Omega, \mu)$ Spaces

Definition 1.39 Let Ω be a set and let $\mathscr{F}$ be a σ-algebra of measurable sets, that is, $\mathscr{F} \subset \mathscr{P}(\Omega)$ is a subset and satisfies the following conditions:

i) $\emptyset \in \mathscr{F}$;
ii) if $\Gamma \subset \mathscr{F}$, then its complement Γ^C belongs to $\mathscr{F}$; and
iii) $\displaystyle\bigcup_{j=1}^{\infty} \Gamma_j \in \mathscr{F}$ whenever $\Gamma_j \in \mathscr{F}$ for all j.

Elements of $\mathscr{F}$ are then called measurable sets.

Definition 1.40 The mapping $\mu : \Omega \mapsto [0, \infty]$ is called a measure, if it satisfies the following conditions:

i) $\mu(\emptyset) = 0$; and
ii) $\displaystyle\mu\left(\bigcup_{j=1}^{\infty} \Gamma_j\right) = \sum_{j=1}^{\infty} \mu(\Gamma_j)$ for any disjoint countable family $(\Gamma_j)_{j \in \mathbb{N}}$ of elements of $\mathscr{F}$.

The σ-algebra $\mathscr{F}$ is said to be σ-finite if there exists a disjoint countable family $(\Gamma_j)_{j \in \mathbb{N}}$ of elements of $\mathscr{F}$ such that $\Omega = \bigcup\limits_{j=1}^{\infty} \Gamma_j$ with $\mu(\Omega_j) < \infty$ for all j.

The space $(\Omega, \mathscr{F}, \mu)$ is called a measure space if Ω is a set, $\mathscr{F}$ is a σ-algebra on Ω, and μ is a measure on Ω.

Let $p \in [1, \infty)$ and let $(\Omega, \mathscr{F}, \mu)$ be a measure space. The space $L^p(\Omega, \mu)$ (also denoted $L^p(\Omega)$) is defined as the collection of all measurable functions $f : \Omega \mapsto \mathbb{C}$ such that

$$\|f\|_{L^p} = \|f\|_p := \left[\int_{\Omega} |f(x)|^p d\mu \right]^{\frac{1}{p}} < \infty.$$

Recall that in $L^p(\Omega, \mu)$, two functions f and g are equal, if they are equal almost everywhere on Ω. Obviously, $(L^p(\Omega, \mu), \|\cdot\|_p)$ is a normed vector space.

Similarly, one defines $L^{\infty}(\Omega, \mu)$ (also denoted $L^{\infty}(\Omega)$) as the set of all measurable functions $f : \Omega \mapsto \mathbb{C}$ such that there exists a constant $M \geq 0$ such that $|f(x)| \leq M$ a.e. $x \in \Omega$. Now define, for all $f \in L^{\infty}(\Omega, \mu)$,

$$\|f\|_{\infty} := \inf \left\{ M : |f(x)| \leq M \text{ a.e. } x \in \Omega \right\}.$$

Obviously, $(L^{\infty}(\Omega, \mu), \|\cdot\|_{\infty})$ is a normed vector space.

In view of the above, $(L^p(\Omega), \|\cdot\|_p)$ is a normed vector space for all $p \in [1, \infty]$.

Example 1.41 Take $\Omega = \mathbb{R}$ and $d\mu = dx$ (Lebesgue measure). Consider the function defined by $f(x) = e^{-|x|}$ for all $x \in \mathbb{R}$. One can easily see that $f \in L^p(\mathbb{R}, dx)$ for all $p \in [1, \infty]$.

Proposition 1.42 *Let (Ω, μ) be a measure space. If $1 \leq p \leq q < \infty$ and if $0 < \mu(\Omega) < \infty$, then*

$$\|f\|_p \leq [\mu(\Omega)]^r \|f\|_q$$

for any measurable function f, where $r = p^{-1} - q^{-1}$.

Proof The proof is left to the reader as an exercise.

Corollary 1.43 *Let (Ω, μ) be a measure space. If $1 \leq p \leq q < \infty$ and if $0 < \mu(\Omega) < \infty$, then the injection*

$$L^q(\Omega, \mu) \hookrightarrow L^p(\Omega, \mu)$$

is continuous.

Proof The proof is left to the reader as an exercise.

Theorem 1.44 (Riesz-Fisher) *The space $(L^p(\Omega, \mu), \|\cdot\|_p)$ is a Banach space for any $1 \leq p \leq \infty$.*

Proof The proof is left to the reader as an exercise.

Theorem 1.45 *Let $p \in [1, \infty]$. If $u \in L^1(\mathbb{R})$ and let $v \in L^p(\mathbb{R}^n)$, then for almost every $x \in \mathbb{R}^n$, the function $y \mapsto u(x - y)v(y)$ is integrable on $\mathbb{R}^n$. And the convolution of u and v defined by*

$$(u * v)(x) := \int_{\mathbb{R}^n} u(x - y)v(y)dy$$

is well defined. Further, $u * v \in L^p(\mathbb{R}^n)$ and

$$\|u * v\|_{L^p} \leq \|u\|_{L^1} \|v\|_{L^p}.$$

Proof See the book by Brézis [31].

More generally,

Theorem 1.46 (Young) *Let $p, q \in [1, \infty]$ such that $r^{-1} = p^{-1} + q^{-1} - 1 \geq 0$. If $u \in L^p(\mathbb{R}^n)$ and let $v \in L^q(\mathbb{R}^n)$, then $u * v \in L^r(\mathbb{R}^n)$ and*

$$\|u * v\|_{L^r} \leq \|u\|_{L^p} \|v\|_{L^q}.$$

Proof See the book by Brézis [31].

Let $\Omega \subset \mathbb{R}^n$ be a subset and let $d\mu = dx$ (Lebesgue measure). Let $L_{loc}^p(\Omega)$ for $1 \leq p < \infty$ stand for the collection of all measurable functions $f : \Omega \mapsto \mathbb{C}$ such that

$$\left(\int_{\Omega'} |f(x)|^p dx \right)^{\frac{1}{p}} < \infty \tag{1.1}$$

for any $\Omega' \subset \Omega$ bounded closed subset.

Clearly, $L_{loc}^p(\Omega)$ is a vector space. Further, $L^p(\Omega)$ is a subspace of $L_{loc}^p(\Omega)$. The natural topology of $L_{loc}^p(\Omega)$ is given as follows: a sequence $(f_n)_{n \in \mathbb{N}} \in L_{loc}^p(\Omega)$ is said to converge to some $f \in L_{loc}^p(\Omega)$ if $\|f_n - f\|_p \to 0$ as $n \to \infty$ in $L^p(\Omega')$ for any $\Omega' \subset \Omega$ bounded closed subset. Although $L_{loc}^p(\Omega)$ equipped with such a type of convergence is a topological vector space, it is not a Banach space.

1.3.4 Sobolev Spaces

If $\alpha = (\alpha_1, \alpha_2, ..., \alpha_n)$ with $\alpha_i \in \mathbb{Z}_+$ for $i = 1, ..., n$, one defines the length $|\alpha|$ of α as follows: $|\alpha| = \alpha_1 + \alpha_2 + ... + \alpha_n$. In this event, the differential operator D^α is defined by

$$D^\alpha = \frac{\partial^{|\alpha|}}{\partial x_1^{\alpha_1} \partial x_2^{\alpha_2} ... \partial x_n^{\alpha_n}}.$$

Definition 1.47 Let $1 \leq p \leq \infty$ and let $k \in \mathbb{N}$. Suppose $\Omega \subset \mathbb{R}^n$ is an open subset. The Sobolev spaces $W^{k,p}(\Omega)$ is the collection of all functions $u : \Omega \mapsto \mathbb{F}$ belonging to the set

$$W^{k,p}(\Omega) := \left\{ u \in L^p(\Omega) : D^\alpha u \in L^p(\Omega) \text{ for } |\alpha| \leq k \right\}. \tag{1.2}$$

If $p = 2$, the Sobolev space $W^{k,2}(\Omega)$ is denoted by $H^k(\Omega)$. The space $W^{k,p}(\Omega)$ is equipped with the norm defined by

$$\|u\|_{k,p} = \left(\sum_{|\alpha| \leq k} \|D^\alpha u\|_p^p \right)^{\frac{1}{p}}, \quad \text{if} \quad 1 \leq p < \infty, \tag{1.3}$$

and

$$\|u\|_{k,\infty} = \max_{|\alpha| \leq k} |D^\alpha u|_\infty \quad \text{if} \quad p = \infty. \tag{1.4}$$

Definition 1.48 If $f : \Omega \mapsto \mathbb{F}$ is a function, then its support denoted $supp(f)$ is defined by $Supp(f) := \overline{\{x \in \Omega : f(x) \neq 0\}}$.

Example 1.49 If $E \subset \mathbb{R}$ is a subset, then the support of the characteristic function χ_E of the set E defined by $\chi_E(x) = 1$ if $x \in E$ and $\chi_E(x) = 0$ if $x \notin E$, is $\overline{E}$ (closure of E).

Definition 1.50 The notation $C_0^\infty(\Omega)$ stands for the collection of all functions $u : \Omega \mapsto \mathbb{R}$ (or $\mathbb{C}$) of class C^∞ with compact support in Ω.

Definition 1.51 The Sobolev space $W_0^{k,p}(\Omega)$ is defined to be the closure of $C_0^\infty(\Omega)$ in the space $W^{k,p}(\Omega)$, that is,

$$W_0^{k,p}(\Omega) = \overline{C_0^\infty(\Omega)}^{W^{k,p}(\Omega)}.$$

If $p = 2$, then the Sobolev space $W_0^{k,2}(\Omega)$ is denoted by $H_0^k(\Omega)$. Further, if $\Omega = \mathbb{R}^n$, then $W_0^{k,p}(\mathbb{R}^n) = W^{k,p}(\mathbb{R}^n)$.

Theorem 1.52 ([47]) *The Sobolev space $W^{k,p}(\Omega)$ is a Banach space.*

Definition 1.53 Let $1 \leq p < \infty$ and let $s = k + \sigma$ where $k \in \mathbb{N}$ and $\sigma \in (0, 1)$. The Sobolev space $W^{s,p}(\Omega)$ is defined by

$$W^{s,p}(\Omega) := \left\{ u \in W^{k,p}(\Omega) : \frac{|D^\alpha u(x) - D^\alpha u(y)|}{\|x - y\|^{\sigma + \frac{n}{p}}} \in L^p(\Omega \times \Omega), \, \forall \alpha, \, |\alpha| = k \right\},$$

whose norm is given by

$$\|u\|_{s,p} = \left(\|u\|_{k,p}^p + \sum_{|\alpha|=k} \int_{\Omega \times \Omega} \frac{|D^\alpha u(x) - D^\alpha u(y)|^p}{\|x-y\|^{p\sigma+n}} dx dy \right)^{1/p}.$$

If $p = 2$, then $W^{s,2}(\Omega)$ is denoted $H^s(\Omega)$. If $s \geq 0$, then we define $W_0^{s,p}(\Omega)$ to be the closure of the space $C_0^\infty(\Omega)$ in the Sobolev space $W^{s,p}(\Omega)$. In particular, $W_0^{s,2}(\Omega)$ is denoted by $H_0^s(\Omega)$.

1.3.5 Embedding Theorems for Sobolev Spaces

In this subsection, we collect some important and useful embedding theorems on Sobolev spaces including the Sobolev–Gagliardo–Nirenberg's Theorem and the Poincaré's Theorem. Although most of the proofs of these theorems are omitted, we will be referring the reader to the appropriate references.

Theorem 1.54 (Sobolev–Gagliardo–Nirenberg) *Let $p \in [1, n)$. Then the following embedding holds,*

$$W^{1,p}(\mathbb{R}^n) \subset L^q(\mathbb{R}^n)$$

where $\dfrac{1}{q} = \dfrac{1}{p} - \dfrac{1}{n}$. Moreover, there exists a constant $C(p, n)$ such that

$$\|u\|_{L^q} \leq C \|\nabla u\|_{L^p}$$

for all $u \in W^{1,p}(\mathbb{R}^n)$, where the gradient ∇u of u is defined by the vector

$$\nabla u = grad\, u = \left(\frac{\partial u}{\partial x_1}, \frac{\partial u}{\partial x_2}, ..., \frac{\partial u}{\partial x_n} \right).$$

Proof See Brézis [32].

Theorem 1.55 *Let $k \in \mathbb{N}$ and let $p \in [1, \infty)$. Then the following continuous embeddings hold,*

$$W^{k,p}(\mathbb{R}^n) \subset L^q(\mathbb{R}^n) \text{ where } \frac{1}{q} = \frac{1}{p} - \frac{k}{n}, \text{ if } \frac{1}{p} - \frac{k}{n} > 0,$$

$$W^{k,p}(\mathbb{R}^n) \subset L^q(\mathbb{R}^n) \text{ for all } q \in [p, \infty), \text{ if } \frac{1}{p} - \frac{k}{n} = 0,$$

$$W^{k,p}(\mathbb{R}^n) \subset L^\infty(\mathbb{R}^n) \text{ if } \frac{1}{p} - \frac{k}{n} < 0.$$

Proof See Brézis [32].

Theorem 1.56 *Let* $\Omega \subset \mathbb{R}^n$ *be an open subset of class* C^1 *with bounded boundary* $\partial\Omega$. *If* $p \in [1, \infty]$, *we have the following continuous embeddings,*

$$W^{1,p}(\Omega) \subset L^q(\Omega) \ where \ \frac{1}{q} = \frac{1}{p} - \frac{k}{n}, \ if \ p < n,$$

$$W^{1,p}(\Omega) \subset L^q(\Omega) \ for \ all \ q \in [p, \infty), \ if \ p = n,$$

$$W^{1,p}(\Omega) \subset L^\infty(\Omega) \ if \ p > n.$$

Proof See Brézis [32].

We also have

Theorem 1.57 (Rellich–Kondrachov) *Let* $\Omega \subset \mathbb{R}^n$ *be a bounded subset and of class* C^1. *If* $p \in [1, \infty]$, *we have the following compact embeddings,*

$$W^{1,p}(\Omega) \subset L^q(\Omega) \ for \ all \ q \in [1, r) \ \frac{1}{r} = \frac{1}{p} - \frac{1}{n}, \ if \ p < n,$$

$$W^{1,p}(\Omega) \subset L^q(\Omega) \ for \ all \ q \in [p, \infty), \ if \ p = n,$$

$$W^{1,p}(\Omega) \subset C(\overline{\Omega}) \ if \ p > n.$$

Proof See Brézis [32].

We also have the so-called Poincaré's inequality.

Theorem 1.58 (Poincaré's Inequality) *Let* $p \in [1, \infty)$ *and let* $\Omega \subset \mathbb{R}^n$ *be an open bounded subset. Then there exists a constant* $C = C(\Omega, p) > 0$ *such that*

$$\|u\|_{L^p} \le C\|\nabla u\|_{L^p}$$

for all $u \in W_0^{1,p}(\Omega)$.

Proof See Brézis [32].

Theorem 1.59 *Let* $\Omega \subset \mathbb{R}^n$ *be an open bounded subset with a smooth boundary* $\partial\Omega$. *Suppose* $0 \le k \le m - 1$. *Then, we have the following embeddings,*

$$W^{m,p}(\Omega) \subset W^{k,q}(\Omega) \ if \ \frac{1}{q} \ge \frac{1}{p} - \frac{m-k}{n}$$

$$W^{m,p}(\Omega) \subset W^{k,q}(\Omega) \ if \ q < \infty, \ and \ \frac{1}{p} = \frac{m-k}{n}.$$

While the second embedding is compact, the first one is compact only if

$$\frac{1}{q} > \frac{1}{p} - \frac{m-k}{n}.$$

Proof See Adams [10].

1.3.6   Bounded Continuous Functions

Let $J \subset \mathbb{R}$ be an interval (possibly unbounded) and let $(\mathbb{X}, \| \cdot \|)$ be a Banach space.

Definition 1.60 Let $BC(J; \mathbb{X})$ denote the space of all bounded continuous functions $f : J \mapsto \mathbb{X}$. The space $BC(J; \mathbb{X})$ will be equipped with the sup-norm defined by

$$\|f\|_\infty := \sup_{t \in J} \|f(t)\|$$

for all $f \in BC(J; \mathbb{X})$.

Theorem 1.61 *The normed vector space $(BC(J; \mathbb{X}), \| \cdot \|_\infty)$ is a Banach space.*

Proof The proof is left to the reader as an exercise.

Proposition 1.62 *If $(f_n)_{n \in \mathbb{N}} \subset BC(J; \mathbb{X})$ such that f_n converges to some f with respect to the sup-norm, then $f \in BC(J; \mathbb{X})$.*

Proof The proof is left to the reader as an exercise.

1.3.7   Hölder Spaces $C^{k,\alpha}(\overline{\Omega})$

Fix once and for all $\alpha \in (0, 1)$. Let $J \subset \mathbb{R}$ be an interval (possibly unbounded) and let $(\mathbb{X}, \| \cdot \|)$ be a Banach space.

The space $C^m(J; \mathbb{X})$ ($m \in \mathbb{N}$) stands for the collection of all m-times continuously differentiable functions from J into $\mathbb{X}$ and let $BC^m(J; \mathbb{X})$ stand for the space,

$$BC^m(J; \mathbb{X}) = \{f \in C^m(J; \mathbb{X}) : f^{(k)} \in BC(J; \mathbb{X}), \ k = 0, 1, ..., m\}$$

equipped with the norm

$$\|f\|_{BC^m(J;\mathbb{X})} := \sum_{k=0}^{m} \|f^{(k)}\|_\infty \ \text{ for all } \ f \in BC^m(J; \mathbb{X}).$$

Definition 1.63 Hölder spaces of continuous functions $C^{0,\alpha}(J; \mathbb{X})$ and $C^{k,\alpha}(J; \mathbb{X})$ for $\alpha \in (0, 1)$ and $k \in \mathbb{N}$ are respectively defined by

$$C^{0,\alpha}(J; \mathbb{X}) = \left\{ f \in BC(J; \mathbb{X}) : [f]_{C^{0,\alpha}(J;\mathbb{X})} = \sup_{t,s \in J, s < t} \frac{\|f(t) - f(s)\|}{(t-s)^\alpha} < \infty \right\}$$

equipped with the norm

$$\|f\|_{C^{0,\alpha}(J;\mathbb{X})} = \|f\|_\infty + [f]_{C^{0,\alpha}(J;\mathbb{X})}, \quad \text{and}$$

$$C^{k,\alpha}(J; \mathbb{X}) = \left\{ f \in BC^k(J; \mathbb{X}) : f^{(k)} \in C^{0,\alpha}(J; \mathbb{X}) \right\}$$

equipped with the norm

$$\|f\|_{C^{k,\alpha}(J;\mathbb{X})} = \|f\|_{BC^k(J;\mathbb{X})} + [f^{(k)}]_{C^{0,\alpha}(J;\mathbb{X})}.$$

Proposition 1.64 *The Hölder spaces $C^{0,\alpha}(J; \mathbb{X})$ and $C^{k,\alpha}(J; \mathbb{X})$ for $\alpha \in (0, 1)$ and $k \in \mathbb{N}$ equipped with their corresponding norms are respectively Banach spaces.*

Proof The proof is left to the reader as an exercise.

Definition 1.65 The Lipschitz space $Lip(J; \mathbb{X})$ is defined by

$$\text{Lip}(J; \mathbb{X}) = \left\{ f \in BC(J; \mathbb{X}) : [f]_{Lip(J;\mathbb{X})} = \sup_{t,s \in J, s < t} \frac{\|f(t) - f(s)\|}{(t-s)} < \infty \right\},$$

and is equipped with the norm defined by

$$\|f\|_{\widetilde{Lip}(J;\mathbb{X})} = \|f\|_\infty + [f]_{Lip(J;\mathbb{X})}.$$

Proposition 1.66 *The Lipschitz space $(Lip(J; \mathbb{X}), \|\cdot\|_{\widetilde{Lip}(J;\mathbb{X})})$ is a Banach space.*

Proof The proof is left to the reader as an exercise.

Definition 1.67 Let $\Omega \subset \mathbb{R}^N$ be an open subset. The Hölder space $C_b^{0,\alpha}(\overline{\Omega})$ consists of all bounded continuous functions $f : \overline{\Omega} \mapsto \mathbb{C}$ such that

$$[f]_{C_b^\alpha(\overline{\Omega})} := \sup_{x \neq y \in \overline{\Omega}} \frac{|f(x) - f(y)|}{\|x - y\|^\alpha} < \infty.$$

Theorem 1.68 *The space* $(C_b^{0,\alpha}(\overline{\Omega}), \| \cdot \|_{C_b^{0,\alpha}(\overline{\Omega})})$ *is a Banach space, where the norm* $\| \cdot \|_{C_b^{0,\alpha}(\overline{\Omega})}$ *is defined by*

$$\|f\|_{C_b^{0,\alpha}(\overline{\Omega})} = \|f\|_\infty + [f]_{C_b^{0,\alpha}(\overline{\Omega})}$$

for all $f \in C_b^{0,\alpha}(\overline{\Omega})$.

Proof The proof is left to the reader as an exercise.

Definition 1.69 Let $k \in \mathbb{N}$. The Hölder space $C_b^{k,\alpha}(\overline{\Omega})$ consists of all functions $f : \overline{\Omega} \mapsto \mathbb{C}$ which are k-times continuously differentiable functions with bounded partial derivatives such that $D^\beta f \in C_b^{0,\alpha}(\overline{\Omega})$ for any multi-index β with $|\beta| = k$.

Theorem 1.70 *The space Hölder space* $(C_b^{k,\alpha}(\overline{\Omega}), \| \cdot \|_{C_b^{k,\alpha}(\overline{\Omega})})$ *is a Banach space, where the norm* $\| \cdot \|_{C_b^{k,\alpha}(\overline{\Omega})}$ *is defined by*

$$\|u\|_{C_b^{k,\alpha}(\overline{\Omega})} = \sum_{|\beta| \leq k} \|D^\beta u\|_\infty + \sum_{|\beta| = k} [D^\beta]_{C_b^{0,\alpha}(\overline{\Omega})}.$$

Proof The proof is left to the reader as an exercise.

Remark 1.71 Note that the subscript "b" in $C_b^{k,\alpha}(\overline{\Omega})$ should be dropped in the case when the domain Ω is bounded. In other words, if Ω is bounded, then $C_b^{k,\alpha}(\overline{\Omega})$ will be denoted $C^{k,\alpha}(\overline{\Omega})$.

1.3.8   Embedding Theorems for Hölder Spaces

Theorem 1.72 *Let* $\Omega \subset \mathbb{R}^n$ *be an open bounded subset with a smooth boundary* $\partial\Omega$. *Suppose* $0 \leq k \leq m - 1$. *Then, we have the following compact embedding,*

$$W^{m,p}(\Omega) \subset C^{k,\alpha}(\overline{\Omega}) \text{ if } \frac{n}{p} < m - (k + \alpha) \text{ with } 0 < \alpha < 1.$$

Proof See Adams [10].

Theorem 1.73 *Let* $\Omega \subset \mathbb{R}^n$ *be an open subset with a smooth boundary* $\partial\Omega$. *Suppose* $m \in \mathbb{Z}_+$ *and* α, β *are given such that* $0 < \alpha < \beta \leq 1$. *Then, we have the following embeddings,*

$$C_b^{m,\alpha}(\overline{\Omega}) \subset C^m(\overline{\Omega}),$$

$$C_b^{m,\beta}(\overline{\Omega}) \subset C_b^{m,\alpha}(\overline{\Omega}).$$

If Ω is bounded, then embeddings,

$$C^{m,\alpha}(\overline{\Omega}) \subset C^m(\overline{\Omega}),$$

$$C^{m,\beta}(\overline{\Omega}) \subset C^{m,\alpha}(\overline{\Omega}).$$

are compact.

Proof See Adams and Fournier [11].

1.3.9   The Dual Space

Let $(\mathbb{X}, \| \cdot \|)$ be a normed vector space over the field $\mathbb{F}$. A functional $\xi : \mathbb{X} \mapsto \mathbb{F}$, $\xi \mapsto \xi(x)$ ($\xi(x)$ is also denoted $\langle \xi, x \rangle$), is said to be linear if it satisfies the following identity, $\xi(\lambda x + \mu y) = \lambda \xi(x) + \mu \xi(y)$ for all $\lambda, \mu \in \mathbb{F}$ and for all $x, y \in \mathbb{X}$.

A functional $\xi : \mathbb{X} \mapsto \mathbb{F}$ is called continuous, if there exists a constant $K \geq 0$ such that

$$|\xi(x)| \leq K\|x\| \tag{1.5}$$

for all $x \in \mathbb{X}$.

The collection of all linear continuous functionals $\xi : \mathbb{X} \mapsto \mathbb{F}$, which we denote by $\mathbb{X}^*$, is called the (topological) dual of $\mathbb{X}$. Clearly, $\mathbb{X}^*$ is a vector space over $\mathbb{F}$ as we can add elements of $\mathbb{X}^*$ up and multiply them by scalars and still get continuous linear functionals. One can endow the dual $\mathbb{X}^*$ of $\mathbb{X}$ with a norm which we denote by $\| \cdot \|_*$ and which is defined as follows: the norm $\|\xi\|_*$ of $\xi \in \mathbb{X}^*$, is the smallest constant K satisfying Eq. (1.5). Consequently,

$$\|\xi\|_* = \sup_{0 \neq x \in \mathbb{X}} \frac{|\langle \xi, x \rangle|}{\|x\|}.$$

Using the definition of the norm $\| \cdot \|_*$, it easily follows that $|\langle \xi, x \rangle| \leq \|\xi\|_* \|x\|$ for all $\xi \in \mathbb{X}^*$ and $x \in \mathbb{X}$. Furthermore,

$$\|\xi\|_* = \sup_{\|x\| \leq 1} |\langle \xi, x \rangle| = \sup_{\|x\|=1} |\langle \xi, x \rangle|.$$

Theorem 1.74 *The normed vector space $(\mathbb{X}^*, \| \cdot \|_*)$ is a Banach space.*

Proof The proof is left to the reader as an exercise.

1.3.10   The Schauder Fixed-Point Theorem

Definition 1.75 A nonempty set S is said to be convex if for all $x, y \in S$ and $\lambda \in [0, 1]$, then $\lambda x + (1 - \lambda) y \in S$.

Theorem 1.76 (The Brouwer Fixed-Point Theorem [76]) *Let* $S \subset \mathbb{F}^n$ *be a nonempty bounded closed convex subset. If the mapping* $T : S \mapsto S$ *is continuous, then T has at least one fixed-point, that is,* $F_T \neq \emptyset$.

Theorem 1.77 (The Schauder Fixed-Point Theorem [76]) *Let* $\mathbb{X}$ *be a Banach space and let* $S \subset \mathbb{X}$ *be a nonempty compact convex subset. If the mapping* $T : S \mapsto S$ *is continuous, then T has at least one fixed-point, that is,* $F_T \neq \emptyset$.

The following concept which measures the "non-compactness" is due to Kuratowski [78].

Definition 1.78 If $D \subset \mathbb{X}$ is a bounded subset, one defines the measure $\alpha(D)$ of non-compactness of D as follows:

$$\alpha(D) := \inf \left\{ d > 0 : \ D \text{ has a finite covering of diameter less than } d \right\}.$$

Definition 1.79 Let $D \subset \mathbb{X}$ be a subset. Suppose that the map $P : D \mapsto \mathbb{X}$ is continuous. The map P is called condensing if for any bounded subset D' of D, $\alpha(D') > 0$ yields

$$\alpha(P(D')) < \alpha(D').$$

We have the following generalization of the Schauder's fixed-point due to Sadovsky.

Theorem 1.80 (The Sadovsky Fixed-Point Theorem [76]) *Let D be a nonempty convex, bounded, and closed subset of a Banach space* $\mathbb{X}$ *and* $F : D \to D$ *be a condensing map. Then F has a fixed point in D.*

Proof The proof makes use of the Schauder's fixed point theorem (Theorem 1.77). Indeed, fix $x \in D$ and let Γ be the set of all closed convex subsets C of D such that $x \in C$ and F maps C into itself.
   Set

$$\Omega = \bigcap_{C \in \Gamma} C \quad \text{and} \quad K = \overline{Conv}\{F(\Omega) \cup \{x\}\},$$

where $Conv$ denotes the convex envelop and $\overline{Conv}$ its closure.
   Using the fact that $x \in \Omega$ and that F maps Ω into itself yields one must have $K \subseteq \Omega$, which, in turn, yields $F(K) \subseteq F(\Omega) \subseteq \Omega$. Now from $x \in K$ it follows

that $K \in \Gamma$. Consequently, $\Omega \subseteq K$ which yields $\Omega = K$, which, in turn, yields $F(K) = F(\Omega) \subseteq K$ and, therefore,

$$\alpha(K) = \alpha(\overline{Conv}\{F(\Omega) \cup \{x\}\}) = \alpha(\{F(\Omega) \cup \{x\}\}) = \alpha(F(\Omega)) = \alpha(F(K)).$$

Using the fact that F is condensing it follows that $\alpha(K) = 0$, which yields Ω is compact. Consequently, F is a continuous function which maps a convex compact set K into itself and so using Schauder theorem it follows that F has a fixed point.

1.3.11 Leray-Schauder Alternative

We will need the following fixed-point theorem in the sequel.

Theorem 1.81 (Leray-Schauder Alternative [61, Theorem 6.5.4]) *Let D be a closed convex subset of a Banach space $\mathbb{X}$ with $0 \in D$. Let $G : D \to D$ be a completely continuous map. Then, either G has a fixed point in D or the set*

$$\left\{ x \in D : x = \lambda G(x),\ 0 < \lambda < 1 \right\}$$

is unbounded.

1.4   Hilbert Spaces

Hilbert spaces play an important role in many areas including mathematical analysis, physics, quantum mechanics, Fourier analysis, partial differential equations, etc. These spaces, which generalize in a natural fashion the Euclidean space, are named after the German mathematician David Hilbert who introduced them in the mathematical literature. Obviously, a Hilbert space is, by design, a Banach space. Some of their basic properties will be discussed in this section and throughout the entire book. For the uncovered material on Hilbert spaces, we refer the interested reader to some of the classical books in functional analysis, i.e., Brézis [31, 32], Conway [38], Eidelman et al. [53], Naylor and Sell [93], etc.

1.4.1   Basic Definitions

In this section, $\mathscr{H}$ stands for a vector space over the field $\mathbb{F}$ where $\mathbb{F} = (\mathbb{R}, |\cdot|)$ or $(\mathbb{C}, |\cdot|)$.

Definition 1.82 A mapping $a : \mathscr{H} \times \mathscr{H} \mapsto \mathbb{F}$ is said to be a sesquilinear form, if

i) the mapping $x \mapsto a(x, y)$ is linear for all $y \in \mathcal{H}$, that is,

$$a(\lambda x + \mu x', y) = \lambda a(x, y) + \mu a(x', y)$$

for all $x, x', y \in \mathcal{H}$; and

ii) the mapping $y \mapsto a(x, y)$ is anti-linear for all $x \in \mathcal{H}$, that is,

$$a(x, \lambda y + \mu y') = \overline{\lambda} a(x, y) + \overline{\mu} a(x, y')$$

for all $x, y, y' \in \mathcal{H}$.

Definition 1.83 An inner product or scalar product $\langle \cdot, \cdot \rangle$ on $\mathcal{H}$ is a sesquilinear form which goes from $\mathcal{H} \times \mathcal{H}$ into $\mathbb{F}$ and satisfies:

i) $\langle x, x \rangle \geq 0$ for all $x \in \mathcal{H}$;

ii) $\langle x, x \rangle = 0$ if and only if $x = 0$; and

iii) $\langle y, x \rangle = \overline{\langle x, y \rangle}$ for all $x, y \in \mathcal{H}$.

Recall that if $\langle \cdot, \cdot \rangle$ is an inner product on $\mathcal{H}$, then the mapping $\| \cdot \| : \mathcal{H} \mapsto \mathbb{R}_+$ defined by $\|x\| := [\langle x, x \rangle]^{\frac{1}{2}}$ for all $x \in \mathcal{H}$ is a norm on $\mathcal{H}$; called the norm deduced from the inner product $\langle \cdot, \cdot \rangle$. Recall also that the norm $\| \cdot \|$ satisfies various properties including the so-called Cauchy-Schwarz inequality and the parallelogram identity given respectively by,

$$\left| \langle x, y \rangle \right| \leq \|x\| \cdot \|y\|, \tag{1.6}$$

and

$$\|x + y\|^2 + \|x - y\|^2 = 2\|x\|^2 + 2\|y\|^2 \tag{1.7}$$

for all $x, y \in \mathcal{H}$.

While the proof of the Cauchy-Schwarz inequality requires more efforts, that of the parallelogram identity is easy and is based upon the following identities:

$$\|x + y\|^2 = \|x\|^2 + 2\Re e \langle x, y \rangle + \|y\|^2$$

and

$$\|x - y\|^2 = \|x\|^2 - 2\Re e \langle x, y \rangle + \|y\|^2$$

for all $x, y \in \mathcal{H}$.

Let $\mathcal{H}$ be a vector space over $\mathbb{F}$ equipped with the inner product given by, $\langle \cdot, \cdot \rangle$. Two vectors $x, y \in \mathcal{H}$ are said to be orthogonal if $\langle x, y \rangle = 0$. If $\langle x, y \rangle = 0$, then the Pythagorean theorem holds in $\mathcal{H}$, that is,

$$\|x + y\|^2 = \|x\|^2 + \|y\|^2.$$

More generally, if $(x_n)_{n \in \mathbb{N}} \subset \mathcal{H}$ is a sequence such that $\langle x_n, x_m \rangle = 0$ for all $n, m \in \mathbb{N}$ with $n \neq m$, then the series $\sum_{n=1}^{\infty} x_n$ converges if and only if $\sum_{n=1}^{\infty} \|x_n\|^2$ does. In that event, we have the following generalized Pythagorean theorem,

$$\left\| \sum_{n=1}^{\infty} x_n \right\|^2 = \sum_{n=1}^{\infty} \left\| x_n \right\|^2.$$

If $M \subset \mathcal{H}$ is a subspace, then its orthogonal $M^{\perp}$ is defined by

$$M^{\perp} := \left\{ x \in \mathcal{H} : \langle x, y \rangle = 0 \text{ for all } y \in M \right\}.$$

Proposition 1.84 *If $M \subset \mathcal{H}$ is a subspace, then its orthogonal $M^{\perp}$ is a closed subspace of $\mathcal{H}$.*

Proof Let $(x_n)_{n \in \mathbb{N}} \subset M^{\perp}$ be a sequence such that $\|x_n - x\| \to 0$ as $n \to \infty$ for some $x \in \mathcal{H}$. Now, using the fact that, $\langle x, y \rangle = \langle x - x_n + x_n, y \rangle = \langle x - x_n, y \rangle + \langle x_n, y \rangle = \langle x - x_n, y \rangle$ for all $y \in M$ and $n \in \mathbb{N}$, it follows, using the Cauchy-Schwarz inequality, that $|\langle x, y \rangle| \leq \|x - x_n\| \cdot \|y\| \to 0$ as $n \to \infty$, which yields $\langle x, y \rangle = 0$, and hence $x \in M^{\perp}$.

Note that in addition to Proposition 1.84, we also have $\mathcal{H}^{\perp} = \{0\}$ and $\{0\}^{\perp} = \mathcal{H}$. Further, if $M \subset N$ where M, N are subspaces of $\mathcal{H}$, then $N^{\perp} \subset M^{\perp}$ and $M \subset (M^{\perp})^{\perp}$ with $(M^{\perp})^{\perp} = M$ if M is closed.

Definition 1.85 The space $\mathcal{H}$ is said to be a Hilbert space if $(\mathcal{H}, \| \cdot \|)$ is complete where $\| \cdot \|$ is the norm deduced from the inner product $\langle \cdot, \cdot \rangle$.

One of the most important properties of Hilbert spaces is that of the projection theorem. It plays an important role in many areas.

Theorem 1.86 *Let $\mathcal{H}$ be a Hilbert space and let $\Sigma \subset \mathcal{H}$ be a nonempty closed convex subset. For each $x \in \mathcal{H}$, there exists a unique point $P_{\Sigma}(x)$ belonging to Σ and called the orthogonal projection of x onto Σ that satisfies the identity*

$$\left\| x - P_{\Sigma}(x) \right\| = \inf_{y \in \Sigma} \left\| x - y \right\|.$$

An immediate consequence of Theorem 1.86 is that if $\mathbb{L} \subset \mathcal{H}$ is a closed subspace, then $\mathcal{H}$ can be written as the direct sum of $\mathbb{L}$ and $\mathbb{L}^{\perp}$ as follows: $\mathcal{H} = \mathbb{L} \oplus \mathbb{L}^{\perp}$. This means that each $x \in \mathcal{H}$ can be uniquely written as $x = (x - P_{\mathbb{L}}(x)) + P_{\mathbb{L}}(x)$ where $P_{\mathbb{L}}(x) \in \mathbb{L}$ and $x - P_{\mathbb{L}}(x) \in \mathbb{L}^{\perp}$. The mapping $P_{\mathbb{L}} : \mathcal{H} \mapsto \mathcal{H}, x \mapsto P_{\mathbb{L}}(x)$, is called the orthogonal projection of $\mathcal{H}$ onto $\mathbb{L}$.

1.4.2 Examples of Hilbert Spaces

Classical examples of Hilbert spaces include, but are not limited to, $\mathbb{R}^n$, $\mathbb{C}^n$, $L^2(\Omega)$, and $\ell^2(\mathbb{N})$, equipped with their natural inner products. Other examples of Hilbert spaces include Sobolev spaces $H^k(\Omega)$. Take $k = 1$, $\Omega = \mathbb{R}$, and consider H^1 defined by

$$H^1(\mathbb{R}) = \{f \in L^2(\mathbb{R}) : f' \in L^2(\mathbb{R})\}.$$

Recall that f' appearing in the definition of $H^1(\mathbb{R})$ is the derivative of f in the sense of distributions. The inner product on $H^1(\mathbb{R})$ and its corresponding norm are given respectively by

$$\langle f, g \rangle = \int_{-\infty}^{\infty} f(t)\overline{g(t)}dt + \int_{-\infty}^{\infty} f'(t)\overline{g'(t)}dt$$

and

$$\|f\|_{H^1(\mathbb{R})} = \left[\int_{-\infty}^{\infty} |f(t)|^2 dt + \int_{-\infty}^{\infty} |f'(t)|^2 dt \right]^{\frac{1}{2}}$$

for all $f, g \in H^1(\mathbb{R})$.

1.5 Exercises

1. Let $(\mathbb{X}, d)$ be a metric space. Show that

$$|d(x, y) - d(y, z)| \leq d(x, z)$$

   for all $x, y, z \in \mathbb{X}$.
2. Let d_0 and d_1 be two metrics on a nonempty set $\mathbb{X}$. Show that mapping d defined by, $d = \alpha d_0 + \beta d_1$ $(d(x, y) = \alpha d_0(x, y) + \beta d_1(x, y)$ for all $x, y \in \mathbb{X})$ where $\alpha, \beta \geq 0$ and $\alpha + \beta = 1$, is a metric on $\mathbb{X}$.
3. Prove Proposition 1.14.
4. Prove Theorem 1.19.
5. Let $a, b > 0$ and let f be the function given for all $x \in \mathbb{R}^n$ by

$$f(x) = \frac{1}{(1 + \|x\|^a)(1 + (\ln \|x\|)^b)},$$

   where $\| \cdot \|$ denotes the Euclidean norm of $\mathbb{R}^n$. Find conditions under which $f \in L^p(\mathbb{R}^n)$.

6. Show that the Lebesgue space $L^p(\Omega)$ is vector space for all $p \in [1, \infty]$.
7. Prove Theorem 1.44.
8. Let $C[a, b]$ the collection of all continuous functions $f : [a, b] \mapsto \mathbb{R}$.

   a) Show that $C[a, b]$ is a vector space over $\mathbb{R}$.
   b) Show that $C[a, b]$ equipped with the norm given, for $p \in [1, \infty)$, by

   $$\|f\|_p := \left(\int_a^b |f(t)|^p dt \right)^{\frac{1}{p}} \text{ for all } f \in C[a, b],$$

   is not a Banach space.

   Hint: Construct a Cauchy sequence in $(C[a, b], \|\cdot\|_p)$ which does not converge.
9. Consider the function $f : \mathbb{R}^n \mapsto \mathbb{R}$ defined by $f(x) = \|x\|^{-\alpha}$ where $\|\cdot\|$ is the Euclidean norm of $\mathbb{R}^n$ and $\alpha \in \mathbb{R}$. Show that $f \in L^1_{loc}(\mathbb{R}^n)$ if and only if $\alpha < n$.
10. Let $1 < p \le \infty$ and let $1 \le q < \infty$ be such that $p^{-1} + q^{-1} = 1$. Show that the (topological) dual of $L^p(\Omega)$ is $L^q(\Omega)$.
11. Let $s \ge 0$ and $1 \le p < \infty$. Suppose q is such that $p^{-1} + q^{-1} = 1$. Show that the (topological) dual of $W_0^{s,p}(\Omega)$ is the Sobolev space $W^{-s,q}(\Omega)$.
12. Prove Theorem 1.64.
13. Prove Theorem 1.66.
14. Prove Theorem 1.68.
15. Prove Theorem 1.70.
16. Prove Theorem 1.74.
17. Prove Theorem 1.86.

1.6   Comments

Some of the basic materials of Sects. 1.3.3, 1.3.4, 1.3.6, and 1.3.10 are taken from the following sources: Adams [10], Adams and Fournier [11], Brézis [31, 32] and Diagana [47]. The material covered in Sects. 1.3.7 and 1.3.9 are partially taken from Bezandry and Diagana [29] and Lunardi [87]. For additional readings upon L^p spaces and Sobolev spaces $W^{k,p}$ we refer the reader to Adams [10], Adams and Fournier [11], and Brézis [32]. For additional readings upon basic functional analysis and real analysis, we refer to Conway [38], Diagana [45], Eidelman et al. [53], Kato [73], Rudin [102], Weidmann [111], and Yosida [112].

The proof of Theorem 1.80 follows Khamsi and Kirk [76, Proof of Theorem 7.12, pages 190–191].

For additional references on metric and normed vector spaces, we refer to the book by Oden and Demkowicz [97]. For additional readings on fixed-point theory, we refer to the book by Khamsi and Kirk [76].

Chapter 2
Operator Theory

In this chapter, the notations $(\mathbb{X}, \| \cdot \|_1)$ and $(\mathbb{Y}, \| \cdot \|_2)$ stand for Banach spaces over the same field $\mathbb{F} = \mathbb{R}$ or $\mathbb{C}$.

2.1 Bounded Linear Operators

2.1.1 Basic Definitions

A transformation $A : \mathbb{X} \mapsto \mathbb{Y}$ is said to be linear if it satisfies the following property,

$$A(tx + sy) = tAx + sAy$$

for all $x, y \in \mathbb{X}$ and $t, s \in \mathbb{F}$.

Definition 2.1 A bounded (or continuous) linear operator is any linear transformation $A : \mathbb{X} \mapsto \mathbb{Y}$ satisfying: there exists a constant $K \geq 0$ such that

$$\|Ax\|_2 \leq K \|x\|_1 \tag{2.1}$$

for all $x \in \mathbb{X}$.

The collection of such bounded linear operators is denoted $B(\mathbb{X}, \mathbb{Y})$. In particular, $B(\mathbb{X}, \mathbb{X})$ is denoted by $B(\mathbb{X})$. If $A \in B(\mathbb{X}, \mathbb{Y})$, then the smallest constant K satisfying Eq. (2.1) is called the norm of A and is denoted by $\|A\|$. Further, it can be shown that the following equalities hold,

$$\|A\| = \sup_{x \in \mathbb{X} \setminus \{0\}} \frac{\|Ax\|_2}{\|x\|_1} = \sup_{\|x\|_1 = 1} \|Ax\|_2 = \sup_{\|x\|_1 \leq 1} \|Ax\|_2.$$

© Springer Nature Switzerland AG 2018

T. Diagana, *Semilinear Evolution Equations and Their Applications*,
https://doi.org/10.1007/978-3-030-00449-1_2

Example 2.2 Consider the normed vector space $\ell^p(\mathbb{N})$ of all p-summable infinite sequences of complex numbers whose norm is given by

$$\|x\|_{\ell^p} = \left(\sum_{k=1}^{\infty} |x_k|^p \right)^{1/p} \quad \text{for all} \quad x = (x_1, x_2, x_3, \ldots) \in \ell^p(\mathbb{N}).$$

Consider the right shift operator S_R on $\ell^p(\mathbb{N})$ is defined by

$$S_R(x_1, x_2, x_3, \ldots) = (0, x_1, x_2, x_3, \ldots) \quad \text{for all} \quad x = (x_1, x_2, \ldots) \in \ell^p(\mathbb{N}).$$

It can be easily shown that $S_R : \ell^p(\mathbb{N}) \mapsto \ell^p(\mathbb{N})$ is a bounded linear operator whose norm is given by $\|S_R\| = 1$. One should point out that one gets similar results as above, if S_R is replaced with the so-called left shift S_L defined by

$$S_L(x_1, x_2, x_3, \ldots) = (x_2, x_3, x_4, \ldots) \quad \text{for all} \quad x = (x_1, x_2, \ldots) \in \ell^p(\mathbb{N}).$$

Indeed, it can be shown that $S_L : \ell^p(\mathbb{N}) \mapsto \ell^p(\mathbb{N})$ is a bounded linear operator whose norm is given by $\|S_L\| = 1$.

Example 2.3 Let $p, q \geq 1$ such that $p^{-1} + q^{-1} = 1$ and let $v = (v_1, v_2, \ldots) \in \ell^q(\mathbb{N})$. Consider the linear operator $T : \ell^p(\mathbb{N}) \mapsto \ell^1(\mathbb{N})$ defined by

$$T(x_1, x_2, x_3, \ldots) = (v_1 x_1, v_2 x_2, v_3 x_3, \ldots) \quad \text{for all} \quad x = (x_1, x_2, \ldots) \in \ell^p(\mathbb{N}).$$

Using the discrete Hölder's inequality it follows that,

$$\|Tx\|_{\ell^1} = \sum_{k=1}^{\infty} |v_k x_k|$$

$$\leq \left(\sum_{k=1}^{\infty} |x_k|^p \right)^{1/p} \cdot \left(\sum_{k=1}^{\infty} |v_k|^q \right)^{1/q}$$

$$= K \|x\|_{\ell^p} \quad \text{for all} \quad x \in \ell^p(\mathbb{N}),$$

where $K = \|v\|_{\ell^q} = \left(\sum_{k=1}^{\infty} |v_k|^q \right)^{1/q}$.

Consequently, T is a well-defined bounded linear operator from $\ell^p(\mathbb{N})$ to $\ell^1(\mathbb{N})$.

It can be shown that $B(\mathbb{X}, \mathbb{Y})$ equipped with the above-mentioned operator norm topology $\| \cdot \|$ is a normed vector space which is complete as stated in the next theorem.

Theorem 2.4 *The space $B(\mathbb{X}, \mathbb{Y})$ equipped with the operator norm topology $\| \cdot \|$ defined previously is a Banach space.*

Proof The proof is left to the reader as an exercise.

If $(\mathbb{L}, \| \cdot \|_3)$ is another Banach space, it can be shown that if $A \in B(\mathbb{X}, \mathbb{Y})$ and $B \in B(\mathbb{Y}, \mathbb{L})$, then $BA \in B(\mathbb{X}, \mathbb{L})$. Moreover,

$$\|BA\| \leq \|B\| \cdot \|A\|.$$

Theorem 2.5 *Let $A : \mathbb{X} \mapsto \mathbb{Y}$ be a linear operator. Then the following statements are equivalent,*

i) *A is continuous;*
ii) *A is continuous at 0;*
iii) *there exists a constant $K \geq 0$ such that $\|Ax\|_2 \leq K\|x\|_1$ for all $x \in \mathbb{X}$.*

Proof The proof is left to the reader as an exercise.

Obviously, both the identity operator $I_{\mathbb{X}} : \mathbb{X} \mapsto \mathbb{X}, x \mapsto I_{\mathbb{X}}(x) = x$ and the zero operator $O_{\mathbb{X}} : \mathbb{X} \mapsto \mathbb{X}, x \mapsto O_{\mathbb{X}}(x) = 0$ are classical elements of $B(\mathbb{X})$. If $A \in B(\mathbb{X}, \mathbb{Y})$, one defines its kernel $N(A)$ and range $R(A)$ as follows,

$$N(A) = \text{Ker}\,(A) = \{x \in \mathbb{X} : Ax = 0\} \text{ and } R(A) = \{Ax : x \in \mathbb{X}\}.$$

2.1.2 Some Classes of Bounded Operators

Definition 2.6 An operator $A \in B(\mathbb{X}, \mathbb{Y})$ is said to be injective if $N(A) = \{0\}$. Similarly, an operator $A \in B(\mathbb{X}, \mathbb{Y})$ is said to be surjective if $R(A) = \mathbb{Y}$.

Definition 2.7 An operator $A \in B(\mathbb{X}, \mathbb{Y})$ is said to be bijective or invertible if $N(A) = \{0\}$ and $R(A) = \mathbb{Y}$. In this event, there exists a unique operator called the inverse of A and denoted $A^{-1} : \mathbb{Y} \mapsto \mathbb{X}$ such that

$$AA^{-1} = I_{\mathbb{Y}} \text{ and } A^{-1}A = I_{\mathbb{X}}.$$

Note that if A^{-1} exists, then it is a bounded linear operator. The collection of all invertible bounded linear operators which go from $\mathbb{X}$ into $\mathbb{Y}$ will be denoted by $\mathscr{O}(\mathbb{X}, \mathbb{Y})$. Further, if $\mathbb{X} = \mathbb{Y}$, then $\mathscr{O}(\mathbb{X}, \mathbb{X})$ is denoted $\mathscr{O}(\mathbb{X})$.

Theorem 2.8 *Let $A \in B(\mathbb{X})$ be such that $\|A\| < 1$. Then $I - A \in \mathscr{O}(\mathbb{X})$ and its inverse $(I - A)^{-1}$ is given by*

$$(I - A)^{-1} = \sum_{k=0}^{\infty} A^k = I + A + A^2 + A^3 + \ldots$$

Proof The proof is left to the reader as an exercise.

Definition 2.9 The rank of an operator $A \in B(\mathbb{X}, \mathbb{Y})$ is defined as dim $R(A)$, the dimension of the vector space $R(A)$. An operator A is said to be an operator of finite rank, if dim $R(A) < \infty$. The collection of all finite rank linear operators from $\mathbb{X}$ into $\mathbb{Y}$ is denoted by $\mathscr{F}(\mathbb{X}, \mathbb{Y})$ with $\mathscr{F}(\mathbb{X}, \mathbb{X}) = \mathscr{F}(\mathbb{X})$.

Example 2.10 Let $\mathbb{X} = \mathbb{Y} = L^2(0, 1)$ be equipped with its L^2-topology $\| \cdot \|_{L^2}$ and let f_i, g_j be functions that belong to $L^2(0, 1)$ for $i, j = 1, 2, ..., N$. Consider the linear operator $A : L^2(0, 1) \mapsto L^2(0, 1)$ defined by

$$(Af)(x) = \int_0^1 K(x, y)f(y)dy$$

where the kernel $K : (0, 1) \times (0, 1) \mapsto \mathbb{F}$ is given by

$$K(x, y) = \sum_{j=1}^N f_j(x)g_j(y).$$

It can be shown that $A \in \mathscr{F}(L^2(0, 1))$.

Definition 2.11 An operator $A \in B(\mathbb{X}, \mathbb{Y})$ is said to be compact if A maps bounded subsets of $\mathbb{X}$ into relatively compact subsets of $\mathbb{Y}$. The collection of all compact linear operators from $\mathbb{X}$ into $\mathbb{Y}$ is denoted by $\mathscr{K}(\mathbb{X}, \mathbb{Y})$.

Definition 2.11 is equivalent to the following definition:

Definition 2.12 An operator $A \in B(\mathbb{X}, \mathbb{Y})$ is compact if for each sequence $(x_n)_{n \in \mathbb{N}}$ in $\mathbb{X}$ such that $\|x_n\|_1 \leq 1$ for each $n \in \mathbb{N}$, the sequence $(Ax_n)_{n \in \mathbb{N}}$ has a subsequence which converges in $\mathbb{Y}$.

Theorem 2.13 *If $A, B \in \mathscr{K}(\mathbb{X}, \mathbb{Y})$ and $\lambda \in \mathbb{F}$, then λA and $A + B$ are compact operators. Moreover, if $C \in B(\mathbb{X})$ and $D \in B(\mathbb{Y})$, then AC and DA are compact operators.*

Proof It is clear that λA is compact for all $\lambda \in \mathbb{F}$.

Let $(x_n) \subset \mathbb{X}$ with $\|x_n\|_1 \leq 1$. Using the fact that A is compact operator it follows that $(Ax_n)_{n \in \mathbb{N}}$ has a convergent subsequence $(Ax_{n_k})_{k \in \mathbb{N}}$ in $\mathbb{Y}$. Similarly, $(Bx_n)_{n \in \mathbb{N}}$ has a convergent subsequence $(Bx_{n_k})_{k \in \mathbb{N}}$. Therefore, $((A + B)x_{n_k})_{k \in \mathbb{N}}$ converges in $\mathbb{Y}$. This implies that $A + B$ is compact.

Let $(y_n)_{n \in \mathbb{N}} \subset \mathbb{X}$ with $\|y_n\| \leq 1$ for each $n \in \mathbb{N}$. consequently, $(Cy_n)_{n \in \mathbb{N}}$ is bounded. Since A is compact, it is clear that $(ACy_n)_{n \in \mathbb{N}}$ has a convergent subsequence in $\mathbb{Y}$. This implies that AC is compact.

Let $(z_n)_{n \in \mathbb{N}} \subset \mathbb{X}$ with $\|z_n\| \leq 1$ for each $n \in \mathbb{N}$. Since A is compact, $(Az_n)_{n \in \mathbb{N}}$ has a convergent subsequence, which we denote $(Az_{n_k})_{k \in \mathbb{N}}$. From the continuity of D it follows that $(DAz_{n_k})_{k \in \mathbb{N}}$ converges. This implies that DA is compact.

Example 2.14 Let $\mathbb{X} = \mathbb{Y} = L^2(0, 1)$ be equipped with its L^2-topology $\| \cdot \|_{L^2}$. Consider the linear operator $A : L^2(0, 1) \mapsto L^2(0, 1)$ defined by

$$(Af)(x) = \int_0^1 K(x, y)f(y)dy$$

where the kernel $K : (0, 1) \times (0, 1) \mapsto \mathbb{F}$ satisfies

$$\int_0^1 \int_0^1 |K(x, y)|^2 dx dy < \infty.$$

It can be shown that $A \in K(L^2(0, 1))$.

2.2   Unbounded Linear Operators

2.2.1   Basic Definitions

Definition 2.15 An unbounded linear operator is any linear transformation A : $D(A) \subset \mathbb{X} \mapsto \mathbb{Y}$ which goes from a subspace $D(A)$ of $\mathbb{X}$ (called the domain of A) into $\mathbb{Y}$.

Definition 2.16 If $A : D(A) \subset \mathbb{X} \mapsto \mathbb{Y}$ is an unbounded linear operator, then its graph is defined by the set,

$$\mathscr{G}(A) = \{(x, Ax) : x \in D(A)\}.$$

Definition 2.17 An unbounded linear operator is said to be closed if $\mathscr{G}(A) \subset \mathbb{X} \times \mathbb{Y}$ is closed. Alternatively, for any $(x_n)_{n \in \mathbb{N}} \subset D(A)$ such that $(x_n)_{n \in \mathbb{N}}$ converges to some $x \in \mathbb{X}$ and $(Ax_n)_{n \in \mathbb{N}}$ converges to some $y \in \mathbb{Y}$, one must have $x \in D(A)$ and $Ax = y$.

Definition 2.18 Let A, B be two linear operators. The operator A is said to be an extension of B if $D(B) \subset D(A)$ and $Ax = Bx$ for all $x \in D(B)$.

Definition 2.19 An unbounded linear operator is said to be closable if it has a closed extension.

Recall that the closure of a given operator $A : D(A) \subset \mathbb{X} \mapsto \mathbb{Y}$ may or may not exist. Obviously, if A is closed, then it is closable. In the same way, any bounded linear operator is closed and hence is closable.

For a given unbounded operator $A : D(A) \subset \mathbb{X} \mapsto \mathbb{X}$, one defines its graph norm as follows:

$$\|x\|_{D(A)} = \|x\| + \|Ax\|$$

for all $x \in D(A)$. It can be easily shown that if A is closed, then $(D(A), \| \cdot \|_{D(A)})$ is a Banach space.

In the rest of this book, the space $(D(A), \| \cdot \|_{D(A)})$ will be denoted by $[D(A)]$.

As in the case of bounded linear operators, if $A : D(A) \subset \mathbb{X} \mapsto \mathbb{X}$ is a unbounded linear operator, then its kernel $N(A)$ and range $R(A)$ are defined as follows,

$$N(A) = \text{Ker}(A) = \{x \in D(A) : Ax = 0\}$$

and

$$R(A) = \{Ax : x \in D(A)\}.$$

2.2.2 Examples of Unbounded Operators

Example 2.20 ([85]) Take $\mathbb{X} = \mathbb{Y} = L^2(\mathbb{R})$ which is endowed with its L^2-topology $\| \cdot \|_{L^2}$. Consider the one-dimensional Laplace differential operator defined by

$$D(A) = H^2(\mathbb{R}) \text{ and } A\psi = -\psi'' \text{ for all } \psi \in H^2(\mathbb{R}).$$

Consider the sequence of functions $\psi_n(x) = e^{-n|x|}$, $n = 1, 2, \dots$. It can be shown that $\psi_n \in D(A) = H^2(\mathbb{R})$ for all $n \in \mathbb{N}$. Moreover,

$$\|\psi_n\|_{L^2}^2 = \int_{-\infty}^{\infty} e^{-2n|x|} dx = \frac{1}{n}$$

and

$$\|A\psi_n\|_{L^2}^2 = \int_{-\infty}^{\infty} n^4 e^{-2n|x|} dx = n^3.$$

Consequently, $\dfrac{\|A\psi_n\|_{L^2}}{\|\psi_n\|_{L^2}} = n \to \infty$ as n goes to ∞ which yields A is an unbounded linear operator on $L^2(\mathbb{R})$.

Example 2.21 ([85]) Take $\mathbb{X} = \mathbb{Y} = L^2(0, 1)$ which is endowed with its L^2-topology $\| \cdot \|_{L^2}$. Consider the first-order derivative operator defined by

$$D(A) = C^1(0, 1) \text{ and } A\psi = \psi' \text{ for all } \psi \in C^1(0, 1),$$

where $C^1(0, 1)$ is the set of all continuously differentiable functions over $(0, 1)$.

Clearly, the sequence of functions defined by: $\phi_n(x) = x^n$, $n = 1, 2, \dots$ belongs to $\in C^1(0, 1)$. Moreover,

$$\|\phi_n\|_{L^2}^2 = \int_0^1 x^{2n} dx = \frac{1}{2n + 1},$$

and

$$\| A\phi_n \|_{L^2}^2 = \int_0^1 n^2 x^{2n-2} \, dx = \frac{n^2}{2n-1}.$$

Now

$$\frac{\| A\phi_n \|_{L^2}}{\| \phi_n \|_{L^2}} = n\sqrt{\frac{2n+1}{2n-1}} \to \infty \text{ as } n \to \infty$$

which yields A is an unbounded linear operator on $L^2(0, 1)$.

2.2.3 The Adjoint Operator in Banach Spaces

Let $A : D(A) \subset \mathbb{X} \mapsto \mathbb{Y}$ be a densely defined ($\overline{D(A)} = \mathbb{X}$) closed unbounded linear operator. Our main objective here is to define the adjoint operator of A. For that, define

$$D(A^*) = \left\{ \xi \in \mathbb{Y}^* : \exists K \geq 0 \text{ such that } |\xi(Ax)| \leq K\|x\|_1 \text{ for all } x \in D(A) \right\}.$$

Clearly, $D(A^*)$ defined above is a subspace of $\mathbb{Y}^*$. Now, define the adjoint operator $A^* : D(A^*) \subset \mathbb{Y}^* \mapsto \mathbb{X}^*$ of A as follows,

$$\xi(Ax) = A^*\xi(x) \text{ for all } x \in D(A), \ \xi \in D(A^*).$$

Theorem 2.22 *Let $A : D(A) \subset \mathbb{X} \mapsto \mathbb{Y}$ be a densely defined closed unbounded linear operator. Then we have,*

  i) $N(A) = R(A^*)^\perp$;
 ii) $N(A^*) = R(A)^\perp$;
iii) $N(A^*)^\perp = \overline{R(A)}$; and
 iv) $\overline{R(A^*)} \subset N(A)^\perp$.

Proof See Brézis [31].

Theorem 2.23 *Let $A : D(A) \subset \mathbb{X} \mapsto \mathbb{Y}$ be a densely defined closed unbounded linear operator. Then the following statements are equivalent,*

  i) $R(A)$ is closed;
 ii) $R(A^*)$ is closed;
iii) $N(A^*)^\perp = R(A)$; and
 iv) $R(A^*) \subset N(A)^\perp$.

Proof See Brézis [31].

If $\mathbb{X} = \mathbb{Y} = \mathcal{H}$ is a Hilbert space equipped with the norm and inner product given respectively by $\| \cdot \|_1$ and $\langle \cdot, \cdot, \rangle$, then the adjoint operator A^* is defined by,

$$\langle Ax, y \rangle = \langle x, A^* y \rangle \text{ for all } x \in D(A), y \in D(A^*)$$

where $D(A^*)$ is defined by

$$D(A^*) = \left\{ y \in \mathcal{H} : \exists K \geq 0 \text{ such that } |\langle Ax, y \rangle| \leq K\|x\|_1 \text{ for all } x \in D(A) \right\}.$$

If $A \in B(\mathcal{H})$, then $D(A) = \mathcal{H}$ and all the previous properties of the adjoint applies to A^*. In addition, $A^* \in B(\mathcal{H})$ and $\|A\| = \|A^*\|$. Further, if $A, B \in B(\mathcal{H})$, then $(A + B)^* = A^* + B^*$; $(AB)^* = B^*A^*$; $(\lambda A)^* = \overline{\lambda}A^*$; $I^* = I$; and $O^* = O$.

Definition 2.24 An unbounded linear operator $A : D(A) \subset \mathcal{H} \mapsto \mathcal{H}$ is said to be self-adjoint if $A^* = A$.

Example 2.25 Let A be the operator defined in the Hilbert space $L^2(\mathbb{R})$ by

$$Af = -i\frac{df}{dx} \text{ for all } f \in D(A) = H^1(\mathbb{R}).$$

It can be shown that $A = A^*$ and so A is a self-adjoint operator.

Definition 2.26 An unbounded linear operator $A : D(A) \subset \mathcal{H} \mapsto \mathcal{H}$ is said to be normal if $A^*A = AA^*$.

Obviously, every self-adjoint operator $A : D(A) \subset \mathcal{H} \mapsto \mathcal{H}$ is normal but the converse is untrue. If A is closed, it can be shown that both AA^* and A^*A are self-adjoint operators.

2.2.4 Spectral Theory

Let $A : D(A) \subset \mathbb{X} \mapsto \mathbb{Y}$ be a closed linear operator. The resolvent set of A denoted $\rho(A)$ is defined by

$$\rho(A) = \left\{ \lambda \in \mathbb{C} : (\lambda I - A)^{-1} \in B(\mathbb{X}) \right\}.$$

Similarly, the spectrum of A denoted $\sigma(A)$ is the complement of the resolvent set $\rho(A)$ in $\mathbb{C}$, that is,

$$\sigma(A) = \mathbb{C} \setminus \rho(A).$$

Although the definition of the spectrum is easy to understand, computing it is in general a very hard task.

Example 2.27 In $L^2(\mathbb{R})^n$, consider the differential operator with constant coefficients defined by

$$A = S(\partial_x) = \sum_{k=0}^{d} a_k \partial_x^k$$

where $a_k \in M(n, \mathbb{C})$ for $k = 0, 1, ..., d$ ($M(n, \mathbb{C})$ being the vector space of $n \times n$-square matrices with complex entries), and

$$S(z) = \sum_{k=0}^{d} a_k z^k \in M(n, \mathbb{C})$$

for each $z \in \mathbb{C}$.

Theorem 2.28 ([25]) *The spectrum of the differential operator A is given by*

$$\sigma(A) = \left\{ \lambda \in \mathbb{C} : \exists \mu \in \mathbb{R}, \ \det[\lambda I_n - S(i\mu)] = 0 \right\}.$$

Example 2.29 Let A be the operator defined in the Hilbert space $L^2(\mathbb{R})$ by

$$Af = -i\frac{df}{dx} \quad \text{for all} \ f \in D(A) = H^1(\mathbb{R}).$$

Then, $\sigma(A) = \mathbb{R}$.

Definition 2.30 The point spectrum $\sigma_p(A)$ of an operator $A : D(A) \subset \mathbb{X} \mapsto \mathbb{X}$ consists of its eigenvalues, that is, all $\lambda \in \mathbb{C}$ such that $\lambda I - A$ is not injective,

$$\sigma_p(A) = \left\{ \lambda \in \mathbb{C} : N(\lambda I - A) \neq \{0\} \right\} = \left\{ \lambda \in \mathbb{C} : \exists u \in \mathbb{X} \setminus \{0\} : \ (\lambda I - A)u = 0 \right\}.$$

Definition 2.31 The residual spectrum $\sigma_r(A)$ of an operator $A : D(A) \subset \mathbb{X} \mapsto \mathbb{X}$ consists of all $\lambda \in \mathbb{C}$ such that $N(\lambda I - A) = \{0\}$ but $R(\lambda I - A)$ is not dense in $\mathbb{X}$,

$$\sigma_r(A) = \left\{ \lambda \in \mathbb{C} : N(\lambda I - A) = \{0\} \ \text{and} \ \overline{R(\lambda I - A)} \neq \mathbb{X} \right\}.$$

Definition 2.32 The continuous spectrum $\sigma_c(A)$ of an operator $A : D(A) \subset \mathbb{X} \mapsto \mathbb{X}$ consists of all $\lambda \in \mathbb{C}$ such that $N(\lambda I - A) = \{0\}$ but $R(\lambda I - A)$ is dense in $\mathbb{X}$,

$$\sigma_c(A) = \left\{ \lambda \in \mathbb{C} : N(\lambda I - A) = \{0\} \ \text{and} \ \overline{R(\lambda I - A)} = \mathbb{X} \right\}.$$

Obviously,

$$\sigma(A) = \sigma_p(A) \cup \sigma_r(A) \cup \sigma_c(A).$$

Another way of defining the spectrum of a linear operator is through both its point spectrum and the so-called essential spectrum.

Definition 2.33 An unbounded linear operator $A : D(A) \subset \mathbb{X} \mapsto \mathbb{X}$ is said to be a Fredholm operator if,

  i) dim $N(A) < \infty$;
 ii) $R(A)$ is closed; and
iii) Codim $R(A) < \infty$.

The collection of those Fredholm operators is denoted $\Phi(\mathbb{X})$.

Classical examples of Fredholm operators include invertible unbounded operators.

Definition 2.34 If $A \in \Phi(\mathbb{X})$, then its index $i(A)$ is defined by

$$i(A) = \dim N(A) - \text{codim } R(A).$$

Definition 2.35 Let $A : D(A) \subset \mathbb{X} \mapsto \mathbb{X}$ be an unbounded linear operator. Then the essential spectrum of A, denoted, $\sigma_{ess}(A)$, is defined by

$$\sigma_{ess}(A) = \Big\{ \lambda \in \mathbb{C} : \lambda I - A \text{ is not a Fredholm operator of index } 0 \Big\}.$$

Proposition 2.36 ([25]) *Let $A : D(A) \subset \mathbb{X} \mapsto \mathbb{X}$ be an unbounded linear operator. Then its spectrum is defined by*

$$\sigma(A) = \sigma_p(A) \cup \sigma_{ess}(A).$$

Proof If $\lambda \in \mathbb{K}$ does not belong to neither $\sigma_p(A)$ nor $\sigma_e(A)$, then $\lambda I - A$ must be injective ($N(\lambda I - A) = \{0\}$) and $R(\lambda I - A)$ is closed with

$$0 = \dim N(\lambda I - A) = \dim \mathbb{X} \setminus R(\lambda I - A).$$

Consequently, $(\lambda I - A)$ must be bijective which yields $\lambda \in \rho(A)$. In view of the above, we have, $\sigma(A) = \sigma_p(A) \cup \sigma_e(A)$.

Example 2.37 Consider the differential operator in Example 2.27. It can be shown (see for instance [25]) that $\sigma_p(A) = \emptyset$ and that

$$\sigma(A) = \sigma_{ess}(A) = \Big\{ \lambda \in \mathbb{C} : \exists \mu \in \mathbb{R}, \ \det(\lambda I_n - S(i\mu)) = 0 \Big\}$$

where $S(z) = \displaystyle\sum_{k=0}^{N} C_k z^k$.

One should stress on the fact that the union $\sigma(A) = \sigma_p(A) \cup \sigma_e(A)$ is not disjoint. It is not hard to find out that the intersection $\sigma_p(A) \cap \sigma_e(A)$ consists of eigenvalues λ of A for which,

(a) either $\dim N(\lambda I - A)$ is not finite,
(b) or $R(\lambda I - A)$ is not closed, and
(c) or $\dim N(A) \neq \dim(\mathbb{X} \setminus R(A))$.

Definition 2.38 Define the continuous spectrum $\sigma_c(A)$ of an unbounded linear operator $A : D(A) \subset \mathbb{X} \mapsto \mathbb{X}$ as follows:

$$\sigma_c(A) := \left\{ \lambda \in \sigma_e(A) \setminus \sigma_p(A) : \overline{R(\lambda I - A)} = \mathbb{X} \right\}.$$

Definition 2.39 Define the residual spectrum $\sigma_r(A)$ of an unbounded linear operator $A : D(A) \subset \mathbb{X} \mapsto \mathbb{X}$ as follows

$$\sigma_r(A) := \left(\sigma_e(A) \setminus \sigma_p(A) \right) \setminus \sigma_c(A).$$

Using the above definitions, through the essential spectrum, one retrieves the usual formula for the spectrum,

$$\sigma(A) = \sigma_p(A) \cup \sigma_c(A) \cup \sigma_r(A).$$

Definition 2.40 The resolvent of A denoted R_λ^A or $R(\lambda, A)$, which maps $\rho(A)$ into $B(\mathbb{X})$, is defined by

$$R_\lambda^A = R(\lambda, A) := (\lambda I - A)^{-1}$$

for all $\lambda \in \rho(A)$.

We have the following properties of the resolvent whose proofs are left to the reader as an exercise.

Proposition 2.41 ([47]) *Let A, B be closed unbounded linear operators on the Banach space $\mathbb{X}$.*

i) *If $\lambda, \mu \in \rho(A)$, then $R(\lambda, A) - R(\mu, A) = (\mu - \lambda)R(\lambda, A)\ R(\mu, A)$. Furthermore, $R(\lambda, A)$ and $R(\mu, A)$ commute.*
ii) *If $D(A) \subset D(B)$, then for all $\lambda \in \rho(A) \cap \rho(B)$ we have*

$$R(\lambda, A) - R(\lambda, B) = R(\lambda, A)(A - B)R(\lambda, B).$$

iii) *If $D(A) = D(B)$, then for all $\lambda \in \rho(A) \cap \rho(B)$ we have*

$$R(\lambda, A) - R(\lambda, B) = R(\lambda, A)(A - B)R(\lambda, B) = R(\lambda, B)(A - B)R(\lambda, A).$$

Definition 2.42 ([47]) A linear operator A is said to have a compact resolvent if $\rho(A) \neq \emptyset$ and $R(\lambda, A) = (\lambda I - A)^{-1}$ is a compact operator for all $\lambda \in \rho(A)$. In particular, A is said to have a compact inverse whether A^{-1} is compact.

2.2.5   Sectorial Operators

Definition 2.43 A linear operator $A : D(A) \subset \mathbb{X} \to \mathbb{X}$ (not necessarily densely defined) is said to be sectorial if there exist constants $\omega \in \mathbb{R}, \theta \in (\frac{\pi}{2}, \pi)$, and $M > 0$ such that the following holds,

i) $\rho(A) \supset S_{\theta,\omega} := \left\{ \lambda \in \mathbb{C} : \lambda \neq \omega, \ |\arg(\lambda - \omega)| < \theta \right\}$, and

ii) $\left\| (\lambda I - A)^{-1} \right\| \leq \dfrac{M}{|\lambda - \omega|}$ for each $\lambda \in S_{\theta,\omega}$.

Recall that the resolvent of a sectorial operator A is nonempty. Therefore, the operator A is closed. Consequently, $[D(A)]$ is a Banach space.

2.2.6   Examples of Sectorial Operators

Example 2.44 Let $p \geq 1$. Take $\mathbb{X} = \mathbb{Y} = L^p(0, 1)$ which is equipped with its natural norm. Consider the linear operator A defined by

$$Au = \frac{d^2 u}{dx^2} \ \text{ for all } \ u \in D(A) = \left\{ u \in W^{2,p}(0, 1) : u(0) = u(1) = 0 \right\}.$$

Then, the linear operator A is sectorial.

Example 2.45 Let $\mathcal{O} \subset \mathbb{R}^n$ be a bounded open subset with C^2 boundary $\partial\mathcal{O}$ and let $\mathbb{X} = \mathbb{Y} = L^2(\mathcal{O})$. Consider the second-order differential operator

$$Au = \Delta u, \ \ \forall u \in D(A) = W^{2,p}(\mathcal{O}) \cap W_0^{1,p}(\mathcal{O}).$$

Then, the linear operator A is sectorial.

Example 2.46 Let $\mathcal{O} \subset \mathbb{R}^N$ be a bounded open subset whose boundary $\partial\mathcal{O}$ is of class C^2 and let $n(x)$ be the outer normal to $\mathcal{O}$ for each $x \in \partial\mathcal{O}$.
Consider

$$A_0 u(x) = \sum_{i,j=1}^{N} a_{ij}(x) \frac{\partial u}{\partial x_i \partial x_j} + \sum_{i=1}^{N} b_i(x) \frac{\partial u}{\partial x_i} + c(x)u(x)$$

where a_{ij} and b_i and c are real, bounded, and continuous on $\overline{\mathcal{O}}$.

Further, we suppose that for each $x \in \overline{\mathscr{O}}$, the matrix $[a_{ij}(x)]_{i,j=1,...,N}$ is symmetric and strictly positive definite, that is,

$$\sum_{i,j=1}^{N} a_{ij}(x)\xi_i\xi_j \geq \omega|\xi|^2 \text{ for all } x \in \overline{\mathscr{O}}, \ \xi \in \mathbb{R}^N.$$

Theorem 2.47 (Agmon [12] and Lunardi et al. [86]) *Let $p > 1$.*

i) *Let $A_p : W^{2,p}(\mathbb{R}^N) \mapsto L^p(\mathbb{R}^N)$ be the linear operator defined by $A_p u = A_0 u$. Then the operator A_p is sectorial in $L^p(\mathbb{R}^N)$ and the domain $D(A_p)$ is dense in $L^p(\mathbb{R}^N)$.*

ii) *Let $\mathscr{O}$ and A_0 defined as above and let A_p be the linear operator defined by*

$$D(A_p) = W^{2,p}(\mathscr{O}) \cap W_0^{1,p}(\mathscr{O}), \quad A_p u = A_0 u.$$

then the linear operator A_p is sectorial in $L^p(\Omega)$. Moreover, $D(A_p)$ is dense in $L^p(\mathscr{O})$.

iii) *Let $\mathscr{O}$ and A_0 defined as above and let A_p be the linear operator defined by*

$$D(A_p) = \{u \in W^{2,p}(\mathscr{O}) : Bu_{|\partial\mathscr{O}} = 0\}, \quad A_p u = A_0 u, \quad u \in D(A_p)$$

where

$$Bu(x) = b_0 u(x) + \sum_{i=1}^{N} b_i(x)\frac{\partial u}{\partial x_i}$$

with the coefficients b_i $(i = 1, ..., N)$ are in $C^1(\overline{\mathscr{O}})$ and the condition

$$\sum_{i=1}^{N} b_i(x)n_i(x) \neq 0 \ x \in \partial\mathscr{O}$$

holds. Then A_p is sectorial in $L^p(\mathscr{O})$ and $D(A_p)$ is dense in $L^p(\mathscr{O})$.

We have the following characterizations of sectorial operators and their perturbations whose proofs can be found in [87, 88].

Proposition 2.48 *Let $A : D(A) \subset \mathbb{X} \mapsto \mathbb{X}$ be a linear operator such that,*

i) *$\{\lambda \in \mathbb{C} : \Re e\lambda \geq \omega\} \subset \rho(A)$; and*
ii) *$\|\lambda(\lambda I - A)^{-1}\| \leq M$ for $\Re e\lambda \geq \omega$ with $\omega \in \mathbb{R}$ and $M > 0$.*

Then the linear operator A is sectorial.

Proposition 2.49 *Let A be a sectorial operator and let B be a linear operator satisfying,*

i) $D(B) \subset D(A)$; and

ii) $\|Bx\| \le a\|x\| + b\|Ax\|$ for all $x \in D(B)$ for some constants $a, b \ge 0$.

There exists $\delta > 0$ such that if $a \in [0, \delta]$, then $A + B$ is sectorial.

Proposition 2.50 *If A is a sectorial operator and if B is a linear operator such that for some constants $\theta \in (0, 1)$ and $C > 0$,*

$$\|Bx\| \le C \left(\|x\| + \|Ax\|\right)^{\theta} \|x\|^{1-\theta}.$$

Then the algebraic sum $A + B$ of A and B is a sectorial linear operator.

2.3 Exercises

Let $(\mathbb{X}, \|\cdot\|)$ be a Banach space over $\mathbb{F}$ and let $(\mathscr{H}, \|\cdot\|)$ be a Hilbert space over $\mathbb{F}$, where $\mathbb{F} = \mathbb{R}$ or $\mathbb{C}$.

1. If $A \in B(\mathbb{X})$, show that its norm $\|A\|$ satisfies,

$$\|A\| = \sup_{\|x\|=1} \|Ax\| = \sup_{\|x\|\le 1} \|Ax\|.$$

2. Prove Theorem 2.4.
3. Prove Theorem 2.5.
4. Prove Theorem 2.8.
5. Show that if $A, B \in \mathscr{O}(\mathbb{X})$, then $AB \in \mathscr{O}(\mathbb{X})$ and that $(AB)^{-1} = B^{-1}A^{-1}$.
6. Show that if $A \in \mathscr{O}(\mathbb{X})$ and if $B \in B(\mathbb{X})$ such that $\|A - B\| < \frac{1}{\|A^{-1}\|}$, then $B \in \mathscr{O}(\mathbb{X})$.
   Hint: Write $B = A[I - A^{-1}(A - B)]$ and use the fact $\|A^{-1}(A - B)\| < 1$.
7. Let $(A_n)_{n\in\mathbb{N}} \subset K(\mathbb{X}, \mathbb{Y})$. Suppose $(A_n)_{n\in\mathbb{N}}$ converges in the operator norm topology to an operator A. Prove that $A \in K(\mathbb{X}, \mathbb{Y})$.
8. Let $A \in B(\mathbb{X})$ and suppose $\lambda \in \sigma(A)$. Show that

$$|\lambda| \le \|A\|.$$

9. The spectral radius $r(A)$ of a linear operator $A : \mathbb{X} \mapsto \mathbb{X}$ is defined by

$$r(A) := \sup_{\lambda\in\sigma(A)} |\lambda|.$$

Show that

$$r(A) \le \|A\|.$$

10. Show that if $A \in B(\mathbb{X})$, then

$$r(A) = \lim_{n \to \infty} \|A^n\|^{\frac{1}{n}}.$$

11. Let $A, B \in B(\mathbb{X})$ such that $AB = BA$. Show that

$$r(AB) \le r(A)r(B).$$

12. Show that the resolvent set $\rho(A)$ of a bounded linear operator A is open.
13. Show that the spectrum $\sigma(A)$ of a bounded linear operator A is a compact set.
14. Prove Theorem 2.23.
15. Prove that for all $A, B \in B(\mathscr{H})$ and $t \in \mathbb{F}$, then

   a. $(A^*)^* = A$;
   b. $(AB)^* = B^*A^*$;
   c. $(tA)^* = \bar{t}A^*$;
   d. $(A + B)^* = A^* + B^*$;
   e. $\|A^*\| = \|A\|$.

16. Prove Theorem 2.28.
17. Let $A : D(A) \subset \mathbb{X} \mapsto \mathbb{X}$ be a sectorial operator. For which values of $\alpha \in \mathbb{C}$ is $A + \alpha I$ sectorial?

2.4 Comments

Most of the materials of this chapter are taken from the following sources: Diagana [47], Bezandry and Diagana [27], and Lunardi [87, 88]. For additional readings on the topics covered in this chapter, we refer the reader to Benzoni [25], Brézis [31, 32], Conway [38, 39], Gohberg et al. [60], Eidelman, Milman and Tsolomitis [53], Kato [73], Rudin [102], Weidmann [111], and Yosida [112].

Chapter 3
Semi-Group of Linear Operators

3.1 Introduction

In this chapter, we collect some of the classical results on strongly continuous semi-groups and evolution families needed in the sequel. All the materials presented here can be found in most of the classical books on semi-groups and evolution families.

While Sect. 3.2 follows Diagana [47], Bezandry and Diagana [27], Chicone and Latushkin [37], Engel and Nagel [55], and Pazy [100], Sect. 3.3 is based upon the following sources: Chicone and Latushkin [37], Lunardi [87], Diagana [47], as well as some articles such as Acquistapace and Terreni [8], Baroun et al. [20], and Schnaubelt [105]. Most of the proofs will not be provided, and therefore, the reader is referred to the above-mentioned references.

3.2 Semi-Group of Operators

3.2.1 Basic Definitions

Definition 3.1 A family of bounded linear operators $(T(t))_{t \in \mathbb{R}_+} : \mathbb{X} \mapsto \mathbb{X}$ is said to be a strongly continuous semi-group or a C_0-semi-group, if

i) $T(0) = I$;
ii) $T(t + s) = T(t)T(s)$ for all $t, s \in \mathbb{R}_+$; and
iii) $\lim_{t \searrow 0} T(t)x = x$ for each $x \in \mathbb{X}$.

© Springer Nature Switzerland AG 2018
T. Diagana, *Semilinear Evolution Equations and Their Applications*,
https://doi.org/10.1007/978-3-030-00449-1_3

If $(T(t))_{t \in \mathbb{R}_+}$ is a C_0-semi-group, then it is the so-called infinitesimal generator, is a linear operator $A : D(A) \subset \mathbb{X} \mapsto \mathbb{X}$ defined by

$$D(A) := \left\{ u \in \mathbb{X} : \lim_{t \searrow 0} \frac{T(t)u - u}{t} \text{ exists} \right\}$$

and

$$Au := \lim_{t \searrow 0} \frac{T(t)u - u}{t}$$

for every $u \in D(A)$.

Example 3.2 Let $p \geq 1$ and let $\mathbb{X} = L^p(\mathbb{R}^n)$ be equipped with its corresponding L^p-norm $\| \cdot \|_{L^p}$.

Definition 3.3 A function $f : \mathbb{R}^n \mapsto \mathbb{C}$ is said to be rapidly decreasing if it is infinitely many times differentiable, that is, $f \in C^\infty(\mathbb{R}^n)$, and

$$\lim_{\|x\| \to \infty} \|x^k D^\alpha f(x)\| = 0$$

for all $k \in \mathbb{N}$ and for all the multi-index $\alpha \in \mathbb{N}^n$.

The Schwartz space is defined by

$$\mathscr{S}(\mathbb{R}^n) = \left\{ f \in C^\infty(\mathbb{R}^n) : f \text{ is rapidly decreasing} \right\}.$$

The Schwartz space is endowed with the family of semi-norms defined by

$$\|f\|_{k,\alpha} = \sup_{x \in \mathbb{R}^n} \|x^k D^\alpha f(x)\|$$

for all $f \in \mathscr{S}(\mathbb{R}^n)$, which makes it a Fréchet space that contains $C_c^\infty(\mathbb{R}^n)$ (class of functions of class C^∞ with compact support) as a dense subspace.

Consider the family of operators in $L^p(\mathbb{R}^n)$ defined by

$$T(t)f(s) := (4\pi t)^{-\frac{n}{2}} \int_{\mathbb{R}^n} e^{\frac{-\|s-r\|^2}{4t}} f(r) dr$$

for all $t > 0$, $s \in \mathbb{R}^n$, and $f \in L^p(\mathbb{R}^n)$.

Proposition 3.4 ([55]) *The family of linear operators $T(t)$ given above, for $t > 0$ and with $T(0) = I$, is a strongly continuous semi-group on $L^p(\mathbb{R}^n)$ whose infinitesimal generator A coincides with the closure of the Laplace operator*

$$\Delta f(x) = \sum_{k=1}^{n} \frac{\partial^2}{\partial x_k^2} f(x_1, x_2, ..., x_n)$$

defined for every f in the Schwartz space $\mathscr{S}(\mathbb{R}^n)$.

Proposition 3.5 ([55]) *If $(T(t))_{t\in\mathbb{R}_+}$ is a C_0-semi-group, then there exists constants $M \geq 1$ and $\omega \in \mathbb{R}$ such that*

$$\|T(t)\| \leq M e^{\omega t} \text{ for all } t \in \mathbb{R}_+.$$

Theorem 3.6 ([100]) *Let $(T(t))_{t\in\mathbb{R}_+}$ be a C_0-semi-group, then the following hold,*

i) For each $x \in \mathbb{X}$,

$$\lim_{h\to 0} \frac{1}{h} \int_t^{t+h} T(s)x\,ds = T(t)x.$$

ii) For each $x \in \mathbb{X}$, $\int_0^t T(s)x\,ds \in D(A)$ and

$$A\left(\int_0^t T(s)x\,ds\right) = T(t)x - x.$$

iii) For all $x \in D(A)$,

$$T(t)x - T(s)x = \int_s^t T(r)Ax\,dr = \int_s^t AT(r)x\,dr.$$

Proposition 3.7 ([47]) *If $(T(t))_{t\in\mathbb{R}_+} : \mathbb{X} \to \mathbb{X}$ is a C_0-semi-group, then the following hold,*

i) the infinitesimal generator A of $T(t)$ is a closed densely defined operator;
ii) the following differential equation holds

$$\frac{d}{dt}T(t)x = AT(t)x = T(t)Ax,$$

holds for every $x \in D(A)$;
iii) for every $x \in \mathbb{X}$, we have $T(t)x = \lim_{s\searrow 0}(\exp(tA_s))x$, with

$$A_s x := \frac{T(s)x - x}{s},$$

where the above convergence is uniform on compact subsets of $\mathbb{R}_+$; and
iv) if $\lambda \in \mathbb{C}$ such that $\Re e\,\lambda > \omega$, then the integral

$$R(\lambda, A)x := (\lambda I - A)^{-1}x = \int_0^\infty e^{-\lambda t} T(t)x\,dt,$$

gives rise to a bounded linear operator $R(\lambda, A)$ on $\mathbb{X}$ whose range is $D(A)$ and satisfies the following identity,

$$(\lambda I - A) \, R(\lambda, A) = R(\lambda, A)(\lambda I - A) = I.$$

If $(T(t))_{t \in \mathbb{R}_+} : \mathbb{X} \to \mathbb{X}$ is a C_0-semi-group, then we know that there exist constants $\omega \in \mathbb{R}$ and $M \geq 1$ such that

$$\|T(t)\| \leq M e^{t\omega}$$

for all $t \in \mathbb{R}_+$.

Now, if $\omega = 0$, then $(T(t))_{t \in \mathbb{R}_+}$ is a C_0-semi-group that is uniformly bounded. If in addition, $M = 1$, then $(T(t))_{t \in \mathbb{R}_+}$ is said to be a C_0-semi-group of contraction. In what follows, we study necessary and sufficient conditions so that A is the infinitesimal generator of a C_0-semi-group of contraction.

Theorem 3.8 (Hille–Yosida) *A linear operator $A : D(A) \to \mathbb{X}$ is the infinitesimal generator of a C_0-semi-group $(T(t))_{t \in \mathbb{R}_+}$ of contraction if and only if,*

 i) *A is a densely defined closed operator; and*
ii) *the resolvent $\rho(A)$ of A contains $[0, \infty)$ and for all $\lambda > 0$,*

$$\left\| (\lambda I - A)^{-1} \right\| \leq \frac{M}{\lambda}. \tag{3.1}$$

For the proof of the Hille-Yosida's theorem, we refer the reader to Pazy [100, Pages 8-9].

3.2.2 Analytic Semi-Groups

Definition 3.9 ([87, Page 34]) A family of bounded linear operators $T(t) : \mathbb{X} \mapsto \mathbb{X}$ satisfying the following conditions (semi-group),

 i) $T(0) = I$;
ii) $T(t + s) = T(t)T(s)$ for all $t, s \geq 0$

is called an analytic semi-group, if $(0, \infty) \mapsto B(\mathbb{X})$, $t \mapsto T(t)$ is analytic.

Definition 3.10 An analytic semi-group $(T(t))_{t \geq 0}$ is said strongly continuous, if the function $[0, \infty) \mapsto \mathbb{X}$, $t \mapsto T(t)x$ is continuous for all $x \in \mathbb{X}$.

It is well known (see for instance Lunardi [87, Page 33]) that if $A : D(A) \subset \mathbb{X} \mapsto \mathbb{X}$ is a sectorial linear operator, then $T(t)$ defined by

$$T(t) = \frac{1}{2i\pi} \int_{w+\gamma_{r,\eta}} e^{\lambda t} (\lambda I - A)^{-1} d\lambda, \quad t > 0 \tag{3.2}$$

is analytic, where $r > 0$, $\eta \in (\frac{\pi}{2}, \pi)$ and $\gamma_{r,\eta}$ is the curve in the complex plane defined by

$$\{\lambda \in \mathbb{C} : |\arg \lambda| = \eta, \ |\lambda| \geq r\} \cup \{\lambda \in \mathbb{C} : |\arg \lambda| \leq \eta, \ |\lambda| = r\},$$

which we assume to be oriented counterclockwise.

Definition 3.11 ([87, Page 34]) If $A : D(A) \subset \mathbb{X} \mapsto \mathbb{X}$ is a sectorial linear operator, then the family of linear operators $\{T(t) : \ t \geq 0\}$ defined in Eq. (3.2) is called the analytic semi-group associated with the operator A.

Proposition 3.12 ([87, Proposition 2.1.1]) *Let A be a sectorial operator and let $T(t)$ be the analytic semi-group associated with it. Then the following hold:*

i) $T(t)u \in D(A^n)$ *for all $t > 0$, $u \in \mathbb{X}$, $n \in \mathbb{N}$. If $u \in D(A^n)$, then*

$$A^n T(t)u = T(t)A^n u, \ t \geq 0;$$

ii) *there exist constants $M_0, M_1, \ldots$ such that*

$$\left\| T(t) \right\| \leq M_0 e^{\omega t}, \ t > 0, \ and$$

$$\left\| t^n (A - \omega I)^n T(t) \right\| \leq M_n e^{\omega t}, \ t > 0; \ and$$

iii) *the mapping $t \to T(t)$ belongs to $C^\infty((0, \infty), B(\mathbb{X}))$ and*

$$\frac{d^n}{dt^n} T(t) = A^n T(t), \ t > 0, \ \forall n \in \mathbb{N}.$$

Proposition 3.13 ([87, Proposition 2.1.9]) *Let $(T(t))_{t>0}$ be a family of bounded linear operators on $\mathbb{X}$ such that $t \mapsto T(t)$ is differentiable, and*

i) $T(t + s) = T(t)T(s)$ *for all $t, s > 0$;*
ii) *there exist $\omega \in \mathbb{R}$, $M_0, M_1 > 0$ such that*

$$\left\| T(t) \right\| \leq M_0 e^{\omega t}, \ \left\| t T'(t) \right\| \leq M_1 e^{\omega t}, \ \forall t > 0;$$

iii) *either: there exists $t > 0$ such that $T(t)$ is one-to-one, or: for every $x \in \mathbb{X}$,*
$\lim_{t \to 0} T(t)x = x$.

Then $t \mapsto T(t)$ is analytic in $(0, \infty)$ with values in $B(\mathbb{X})$, and there exists a unique sectorial operator $A : D(A) \subset \mathbb{X} \to \mathbb{X}$ such that $(T(t))_{t \geq 0}$ is the semi-group associated with A.

3.2.3   Hyperbolic Semi-Groups

Definition 3.14 ([55]) A strongly continuous semi-group $(T(t))_{\in \mathbb{R}_+}$ is called

i) Uniformly exponentially stable, if there exists $\varepsilon > 0$ such that

$$\lim_{t \to \infty} e^{\varepsilon t} \|T(t)\| = 0.$$

ii) Uniformly stable, if

$$\lim_{t \to \infty} \|T(t)\| = 0.$$

iii) Strongly stable, if

$$\lim_{t \to \infty} \|T(t)x\| = 0 \text{ for all } x \in \mathbb{X}.$$

Definition 3.15 A semi-group $(T(t))_{t \in \mathbb{R}_+} : \mathbb{X} \mapsto \mathbb{X}$ is said to be hyperbolic if the Banach space $\mathbb{X}$ can be decomposed as a direct sum

$$\mathbb{X} = \mathbb{X}_s \oplus \mathbb{X}_u$$

where $\mathbb{X}_s$ (stable) and $\mathbb{X}_u$ (unstable) are two $T(t)$-invariant closed subspaces such that $T_s(t)$ the restriction of $T(t)$ to $\mathbb{X}_s$ and $T_u(t)$ the restriction of $T(t)$ to $\mathbb{X}_u$ satisfy the following properties,

i) $(T_s(t))_{t \in \mathbb{R}_+}$ is a semi-group that is uniformly exponentially stable on $\mathbb{X}_s$; and
ii) $(T_u(t))_{t \in \mathbb{R}_+}$ is a semi-group that is invertible on $\mathbb{X}_u$ and its inverse $((T_u)^{-1}(t))_{t \in \mathbb{R}_+}$ is uniformly exponentially stable on $\mathbb{X}_u$.

Remark 3.16 It can be shown that a C_0-semi-group $(T(t))_{t \in \mathbb{R}_+}$ is hyperbolic if and only if there exist a projection $P : \mathbb{X} \mapsto \mathbb{X}$ and some constants $M, \delta > 0$ such that $T(t)P = PT(t)$ for all $t \in \mathbb{R}_+$, $T(t)(N(P)) = N(P)$, and

i) $\|T(t)x\| \leq M e^{-\delta t} \|x\|$ for all $t \in \mathbb{R}_+$ and $x \in R(P)$; and
ii) $\|T(t)x\| \geq \frac{1}{M} e^{\delta t} \|x\|$ for all $t \in \mathbb{R}_+$ and $x \in N(P)$.

Proposition 3.17 ([55, Proposition 3.1.3]) *For a C_0-semi-group $(T(t))_{t \in \mathbb{R}_+}$, the following statements are equivalent,*

i) $(T(t))_{t \in \mathbb{R}_+}$ *is hyperbolic.*
ii) $\sigma(T(t)) \cap \mathbb{S}^1 = \emptyset$ *for one (for all)* $t > 0$, *where* $\mathbb{S}^1 = \{z \in \mathbb{C} : |z| = 1\}$.

Definition 3.18 A C_0-semi-group $(T(t))_{t \in \mathbb{R}_+}$ is said to have the circular spectral mapping theorem, if

$$\mathbb{S}^1 . T(t) \setminus \{0\} = \mathbb{S}^1 e^{t\sigma(A)} \text{ for one (for all) } t > 0,$$

where A is the infinitesimal generator of $(T(t))_{t \in \mathbb{R}_+}$.

Proposition 3.19 ([55, Theorem 3.1.5]) *Let $(T(t))_{t\in\mathbb{R}_+}$ be a C_0-semi-group that has the circular spectral mapping theorem. Then, the following statements are equivalent,*

   i) $(T(t))_{t\in\mathbb{R}_+}$ *is hyperbolic.*
  ii) $\sigma(T(t)) \cap \mathbb{S}^1 = \emptyset$ *for one (for all) $t > 0$.*
 iii) $\sigma(A) \cap i\mathbb{R} = \emptyset$.

3.3 Evolution Families

This section introduces evolution families which play a central role when it comes to dealing with nonautonomous differential equations on Banach spaces. For more on evolution equations and related issues, we refer to the following books: Chicone and Latushkin [78] and Lunardi [87].

3.3.1 Basic Definitions

Let $J \subset \mathbb{R}$ be an interval (possibly unbounded).

Definition 3.20 A family of bounded linear operators $\mathcal{U} = \{U(t, s) : t, s \in J, \ t \geq s\}$ on $\mathbb{X}$ is called an evolution family (also called "evolution systems," "evolution operators," "evolution processes," "propagators," or "fundamental solutions") if the following hold,

   i) $U(t, s)U(s, r) = U(t, r)$ for $t, s, r \in J$ such that $t \geq s \geq r$;
  ii) $U(t, t) = I$ for all $t \in J$.

    The evolution family $\mathcal{U}$ is called strongly continuous if, for each $x \in \mathbb{X}$, the function, $J \times J \mapsto \mathbb{X}, (t, s) \mapsto U(t, s)x$, is continuous for all $s, t \in J$ with $t \geq s$.

Example 3.21 If $(T(t))_{t\in\mathbb{R}_+}$ is a C_0-semi-group, then U defined by $U(t, s) = T(t - s)$ for all $t, s \in J = \mathbb{R}_+$ with $t \geq s$, is an evolution family that is strongly continuous.

Definition 3.22 The exponential growth bound $\omega(\mathcal{U})$ of an evolution family $\mathcal{U} = \{U(t, s) : t, s \in J, \ t \geq s\}$ on $\mathbb{X}$ is defined by

$$\omega(\mathcal{U}) := \inf\left\{\sigma \in \mathbb{R} : \exists M_\sigma \geq 1, \ \|U(t, s)\| \leq M_\sigma e^{\sigma(t-s)} \text{ for all } t, s \in J, \ t \geq s\right\}.$$

Definition 3.23 An evolution family $\mathcal{U} = \{U(t, s) : t, s \in \mathbb{R}, \ t \geq s\}$ defined on $\mathbb{X}$ is called exponentially bounded if $\omega(\mathcal{U}) < \infty$ and $\mathcal{U}$ is exponentially stable if $\omega(\mathcal{U}) < 0$.

Obviously, the evolution family defined by $U(t, s) = T(t - s)$ where $(T(t))_{t \geq 0}$ is an exponentially stable C_0-semi-group, and is an example of an evolution family which is exponentially stable.

3.3.2 Acquistapace–Terreni Conditions

Definition 3.24 A family of linear operators $(A(t))_{t \in \mathbb{R}}$ (not necessarily densely defined) is said to satisfy the Acquistapace–Terreni conditions whether there exist $\omega \in \mathbb{R}$ and the constants $\theta \in (\frac{\pi}{2}, \pi)$, $L, K \geq 0$, and $\mu, \nu \in (0, 1]$ with $\mu + \nu > 1$ such that

$$\Sigma_\theta \cup \{0\} \subseteq \rho(A(t) - \omega I) \ni \lambda, \quad \|R(\lambda, A(t) - \omega I)\| \leq \frac{K}{1 + |\lambda|} \qquad (3.3)$$

and

$$\|(A(t) - \omega I) R(\omega, A(t) - \omega I) \left[R(\omega, A(t)) - R(\omega, A(s)) \right] \| \leq K |t - s|^\mu |\lambda|^{-\nu} \qquad (3.4)$$

for $t, s \in \mathbb{R}$, $\lambda \in \Sigma_\theta := \{\lambda \in \mathbb{C} \setminus \{0\} : |\arg \lambda| \leq \theta\}$.

Recall that Acquistapace–Terreni conditions were introduced by Acquistapace and Terreni in [8, 9] for $\omega = 0$. Among other things, Eqs. (3.3) and (3.4) yield the existence of an evolution family

$$\mathscr{U} = \{U(t, s) : t, s \in \mathbb{R}, \ t \geq s\}$$

associated with $A(t)$ such that for all $t, s \in \mathbb{R}$ with $t > s$, then $(t, s) \mapsto U(t, s)$, $\mathbb{R} \times \mathbb{R} \mapsto B(\mathbb{X})$ is strongly continuous and continuously differentiable in $t \in \mathbb{R}$, $U(t, s)\mathbb{X} \subseteq D(A(t))$ for all $t, s \in \mathbb{R}$,

$$\partial_t U(t, s) = A(t)U(t, s),$$

(a)

$$\|A(t)^k U(t, s)\| \leq C \ (t - s)^{-k} \qquad (3.5)$$

for $0 < t - s \leq 1$, $k = 0, 1$; and

(b) $\partial_s^+ U(t, s)x = -U(t, s)A(s)x$ for $t > s$ and $x \in D(A(s))$ with $A(s)x \in D(A(s))$.

Remark 3.25 Recall that if $D(A(t)) = D$ is constant in $t \in \mathbb{R}$, then Eq. (3.4) (see for instance [100]) can be replaced with the identity: there exist constants L and $0 < \mu \leq 1$ such that

$$\| (A(t) - A(s)) R(\omega, A(r))\| \leq L|t - s|^\mu, \ s, t, r \in \mathbb{R}. \qquad (3.6)$$

Definition 3.26 An evolution family $\mathscr{U} = \{U(t, s) : t, s \in \mathbb{R}, \ t \geq s\} \subset B(\mathbb{X})$ is said to have an *exponential dichotomy (or is hyperbolic)* if there are projections $P(t)$ that are uniformly bounded and strongly continuous in t and constants $\delta > 0$ and $N \geq 1$ such that

i) $U(t, s)P(s) = P(t)U(t, s)$ for all $t \geq s$;

ii) $U(t, s) : Q(s)\mathbb{X} \to Q(t)\mathbb{X}$ is invertible with inverse $\widetilde{U}(s, t)$; and

iii) $\|U(t, s)P(s)\| \leq Ne^{-\delta(t-s)}$ and $\|\widetilde{U}(s, t)Q(t)\| \leq Ne^{-\delta(t-s)}$ for $t \geq s$ and $t, s \in \mathbb{R}$, where $Q(t) = I - P(t)$.

3.3.2.1 Estimates for $U(t, s)$

Fix once and for all $\alpha \in (0, 1)$ let $A : D(A) \subset X \mapsto X$ be a sectorial operator. We will be using the following real interpolation space in the sequel:

$$\mathbb{X}_\alpha^A := \left\{ x \in \mathbb{X} : \|x\|_\alpha^A := \sup_{r>0} \|r^\alpha(A - \zeta)R(r, A - \zeta)x\| < \infty \right\}.$$

Clearly, $(\mathbb{X}_\alpha^A, \|\cdot\|_\alpha^A)$ is a Banach space.

We also define

$$\mathbb{X}_0^A := \mathbb{X}, \ \|x\|_0^A := \|x\|, \ \mathbb{X}_1^A := D(A), \ \hat{\mathbb{X}}^A := \overline{D(A)}, \ \text{and} \ \|x\|_1^A := \|(\zeta - A)x\|.$$

Obviously, the following continuous embedding holds,

$$D(A) \hookrightarrow \mathbb{X}_\beta^A \hookrightarrow D((\zeta - A)^\alpha) \hookrightarrow \mathbb{X}_\alpha^A \hookrightarrow \hat{\mathbb{X}}^A \subset \mathbb{X}, \tag{3.7}$$

for all $0 < \alpha < \beta < 1$.

Similarly,

$$\mathbb{X}_\beta^A \hookrightarrow \overline{D(A)}^{\|\cdot\|_\alpha^A} \tag{3.8}$$

for $0 < \alpha < \beta < 1$.

If the family of linear operators $A(t)$ for $t \in \mathbb{R}$ satisfies Acquistapace–Terreni conditions, we let

$$\mathbb{X}_\alpha^t := \mathbb{X}_\alpha^{A(t)}, \quad \hat{\mathbb{X}}^t := \hat{\mathbb{X}}^{A(t)}$$

for $0 \leq \alpha \leq 1$ and $t \in \mathbb{R}$.

Recall that the above interpolation spaces are of class $\mathscr{J}_\alpha$ and hence there is a constant $l(\alpha)$ such that

$$\|y\|_\alpha^t \leq l(\alpha)\|y\|^{1-\alpha}\|A(t)y\|^\alpha, \quad y \in D(A(t)). \tag{3.9}$$

Proposition 3.27 ([20, 47]) *For* $x \in \mathbb{X}$, $0 \le \alpha \le 1$ *and all* $t > s$, *the following hold,*

(a) *There is a constant* $c(\alpha)$, *such that*

$$\|U(t,s)P(s)x\|_\alpha^t \le c(\alpha)e^{-\frac{\delta}{2}(t-s)}(t-s)^{-\alpha}\|x\|. \tag{3.10}$$

(b) *There is a constant* $m(\alpha)$, *such that*

$$\|\widetilde{U}(s,t)Q(t)x\|_\alpha^s \le m(\alpha)e^{-\delta(t-s)}\|x\|. \tag{3.11}$$

3.4 Exercises

1. Let $T(t)$ be the family of linear operators on $L^2(\mathbb{R}^n)$ defined for all $t > 0$ by

$$T(t)f(s) := (4\pi t)^{-\frac{n}{2}} \int_{\mathbb{R}^n} e^{\frac{-\|s-r\|^2}{4t}} f(r)dr$$

for all $s \in \mathbb{R}^n$, and $f \in L^2(\mathbb{R}^n)$, and by setting $T(0) = I$.

   Show that $(T(t))_{t\ge 0}$ defined above is a C_0-semi-group whose generator is the Laplace operator Δ.

2. Let $BUC(\mathbb{R})$ denote the collection of all real-valued uniformly continuous bounded functions equipped with the sup norm defined by $\|f\|_\infty = \sup_{t\in\mathbb{R}} |f(t)|$. Let $T(t)$ be the family of linear operators on $BUC(\mathbb{R})$ defined by

$$T(t)f(s) = f(t+s)$$

for all $t, s \in \mathbb{R}$, and $f \in BUC(\mathbb{R})$.

   Show that $(T(t))_{t\ge 0}$ defined above is a C_0-semi-group whose generator is the linear operator defined by

$$D(A) = \{f \in BUC(\mathbb{R}) : f' \in BUC(\mathbb{R})\}, \text{ and } Af = f'$$

for all $f \in D(A)$.

3. Prove Proposition 3.5.
4. Prove Proposition 3.7.
5. Let $(T(t))_{t\in\mathbb{R}_+}$ be a C_0-semi-group $(T(t))_{t\in\mathbb{R}_+}$. Show that

   (a) For each $x \in \mathbb{X}$,

$$\lim_{h\to 0} \frac{1}{h} \int_t^{t+h} T(s)xds = T(t)x.$$

(b) For each $x \in \mathbb{X}$, $\int_0^t T(s)x\,ds \in D(A)$ and

$$A\left(\int_0^t T(s)x\,ds\right) = T(t)x - x.$$

(c) For all $x \in D(A)$,

$$T(t)x - T(s)x = \int_s^t T(r)Ax\,dr = \int_s^t AT(r)x\,dr.$$

6. Prove Theorem 3.8.
7. Prove Proposition 3.17.
8. Prove Proposition 3.19.
9. Let A be the linear operator defined on $C([0, 1])$ by $Af = af'' + bf$ for all
   $f \in D(A) = \{g \in C^2([0, 1]) : g(0) = g(1) = 0\}$ where $a > 0$ and $b \neq 0$ are
   constant real numbers.

   (a) Show that $\sigma(A) = \{b - n^2\pi^2\sqrt{a} : n \in \mathbb{N}\}$.
   (b) Show that A is the infinitesimal generator of an analytic semi-group
       $(T(t))_{t \geq 0}$.
   (c) Show that $(T(t))_{t \geq 0}$ is not strongly continuous at $t = 0$.
   (d) Find conditions on a and b so that the semi-group $(T(t))_{t \geq 0}$ is hyperbolic,
       that is, $\sigma(A) \cap i\mathbb{R} = \emptyset$.

10. Let $A : D(A) \subset \mathbb{X} \mapsto \mathbb{X}$ be a linear operator on a Banach space $\mathbb{X}$ and let
    $a : \mathbb{R} \mapsto \mathbb{R}$ be a bounded continuous function satisfying

    $$\inf_{t \in \mathbb{R}} a(t) = a_0 > 0.$$

    Suppose that A is the infinitesimal generator of a C_0-semi-group $(T(t))_{t \in \mathbb{R}_+}$.

    (a) Show that $\mathscr{U} = \{U(t, s) : t, s \in \mathbb{R}, \ t \geq s\}$ defined on $\mathbb{X}$ by

    $$U(t, s) = T\left(\int_s^t a(r)dr\right)$$

    is an evolution family.
    (b) Show that if $(T(t))_{t \in \mathbb{R}_+}$ is exponentially stable, then so is $\mathscr{U}$.
    (c) Find the generator associated with the evolution family $\mathscr{U}$.

3.5   Comments

The material on semi-groups is taken from various sources. Among them are
Diagana [47], Bezandry and Diagana [27], Pazy [100], Chicone and Latushkin [37],
Engel and Nagel [55], and Lunardi [87]. Section 3.3 is mainly based upon the
following books: Chicone and Latushkin [37], Lunardi [87], Diagana [47], Bezandry
and Diagana [27], as well as some articles including Acquistapace and Terreni [8],
Baroun et al. [20], and Schnaubelt [105].

Most of the proofs are omitted, and therefore, the reader is referred to the above-
mentioned references for proofs and additional material on the different topics
discussed in this chapter.

Chapter 4
Almost Periodic Functions and Sequences

In this chapter we review basic properties of almost periodic functions as well as their discrete counterparts needed in the sequel. It is mainly based upon various sources including Corduneanu [41], Diagana [47], Halanay and Rasvan [63], and Levitan and Zhikov [82]. For the sake of completeness and clarity, the proofs of some of those nontrivial results are given in great details. Beyond the above-mentioned references, additional readings on the theory of almost periodic functions and related topics can for instance be found in Besicovitch [26], Bohr [29], Corduneanu [40], Pankov [99], Bezandry and Diagana [27], Fink [58], Henry [65], Amerio and Prouse [13], Levitan [81], N'Guérékata [94, 95], Zhang [114], Hino et al. [70], Diagana et al. [49], Diagana [45], Diagana and Pennequin [48], Fan [56], etc.

4.1 Introduction

In this chapter we study almost periodic functions and almost periodic sequences. Basically, an almost periodic function $f : \mathbb{R} \mapsto \mathbb{X}$ is any continuous function that is the uniform limit of a trigonometric polynomial. In particular, any continuous periodic function and more generally any trigonometric polynomial is almost periodic. Thus if $f, g : \mathbb{R} \mapsto \mathbb{X}$ are almost periodic functions and if $\lambda \in \mathbb{F}$, then both $f + g$ and λf are almost periodic functions. Further, it can be shown that the space of almost periodic functions equipped with the sup norm $\| \cdot \|_\infty$ is a Banach space.

Almost periodic functions play a crucial role in many fields including mathematical analysis, signal processing, harmonic analysis, physics, dynamical systems, etc. These functions were introduced in the mathematical literature around 1924–1926 with the landmark work of the Danish mathematician Bohr [28]. A decade after

© Springer Nature Switzerland AG 2018
T. Diagana, *Semilinear Evolution Equations and Their Applications*,
https://doi.org/10.1007/978-3-030-00449-1_4

their introduction in the literature, various other significant contributions on these functions were made by, among others, S. Bochner, H. Weyl, A. Besicovitch, J. von Neumann, V. V. Stepanov, and many other people.

4.2   Almost Periodic Functions

4.2.1   Basic Definitions

If $f : \mathbb{R} \mapsto \mathbb{X}$ is a function, then the notation $\mathscr{R}_s f$ stands for the function defined by

$$\mathscr{R}_s f(t) = f(t + s)$$

for all $s, t \in \mathbb{R}$.

Definition 4.1 A set $\mathscr{P} \subset \mathbb{R}$ is said to be relatively dense if there exists $\ell > 0$ such that

$$\mathscr{P} \cap (\alpha, \alpha + \ell) \neq \emptyset$$

for all $\alpha \in \mathbb{R}$.

Definition 4.2 A real number τ is said to be an ε-period of a function $f : \mathbb{R} \mapsto \mathbb{X}$ if

$$\sup_{t \in \mathbb{R}} \|\mathscr{R}_\tau f(t) - f(t)\| < \varepsilon.$$

The collection of all the ε-periods for the function f is denoted by $\mathscr{P}(\varepsilon, f)$.

Definition 4.3 (Bohr) A function $f \in C(\mathbb{R}, \mathbb{X})$ is called (Bohr) almost periodic if for $\varepsilon > 0$, $\mathscr{P}(\varepsilon, f)$ is relatively dense. Equivalently, there exists $\ell(\varepsilon) > 0$ such that every interval $(\alpha, \alpha + \ell(\varepsilon))$ of length $\ell(\varepsilon)$ contains a $\tau = \tau(\varepsilon) \in \mathscr{P}(\varepsilon, f)$. The collection of all almost periodic functions $f : \mathbb{R} \mapsto \mathbb{X}$ will be denoted by $AP(\mathbb{X})$ or $AP(\mathbb{R}, \mathbb{X})$.

Standard examples of almost periodic functions include trigonometric polynomials, that is, any function $P_N : \mathbb{R} \mapsto \mathbb{X}$ of the form

$$P_N(t) = \sum_{k=1}^{N} a_k e^{i\lambda_k t} \tag{4.1}$$

where $\lambda_k \in \mathbb{R}$ and $a_k \in \mathbb{X}$ for $k = 1, \ldots, N$.

Every continuous periodic function is almost periodic but the converse is untrue. Indeed, let $\alpha, \beta \in \mathbb{R}$ be such that $\alpha\beta^{-1} \in \mathbb{R} \setminus \mathbb{Q}$ and consider the function defined by $f(t) = \sin \alpha t + \sin \beta t$. It is clear that f is not periodic although it is the sum of two periodic functions. Further, it can be shown that f in fact is almost periodic.

Theorem 4.4 *If $f, g : \mathbb{R} \mapsto \mathbb{X}$ are almost periodic functions, then the following properties hold,*

i) *The function f is uniformly continuous.*
ii) *The range $R(f) = \{f(t) : t \in \mathbb{R}\}$ is relatively compact in $\mathbb{X}$.*
iii) *If g is $\mathbb{F}$-valued such that $0 < M \le |g(t)|$ for each $t \in \mathbb{R}$ for some constant M, then the quotient function $t \mapsto fg^{-1}(t)$ is almost periodic.*
iv) *Let $(f_n)_{n \in \mathbb{N}}$ be a sequence of almost periodic functions such that there exists $h \in C(\mathbb{R}, \mathbb{X})$ with $\|f_n - h\|_\infty \to 0$ as $n \to \infty$. Then h is almost periodic.*

Proof We only provide the proofs of items i) and ii). For the proofs of iii) and iv), we refer the reader to the books by Corduneanu [40, 41].

i) From $f : \mathbb{R} \mapsto \mathbb{X}$ is almost periodic, it follows that for each $\varepsilon > 0$ there exists $\ell(\varepsilon) > 0$ such that every interval of length $\ell(\varepsilon)$ contains a number τ with the property

$$\|f(t + \tau) - f(t)\| < \frac{\varepsilon}{3}$$

for all $t \in \mathbb{R}$.

Using the fact that f is uniform continuity on compact intervals of $\mathbb{R}$, let $\delta \in (0, 1)$ such that

$$\|f(t) - f(s)\| \le \varepsilon$$

for $0 \le t, s \le \ell(\varepsilon) + 1$ and $|t - s| \le \delta$.

Let $t, s \in \mathbb{R}$ be such that $|t - s| \le \delta$. Let τ be such that $0 < t + \tau, s + \tau < \ell(\varepsilon) + 1$.

Now

$$\|f(t) - f(s)\| \le \|f(t) - f(t + \tau)\| + \|f(t + \tau) - f(s - \tau)\|$$
$$+ \|f(s + \tau) - f(s)\|$$
$$< \varepsilon.$$

ii) From $f \in AP(\mathbb{X})$, it follows that, for each $\varepsilon > 0$, there exists $\ell(\varepsilon) > 0$ such that every interval of length $\ell(\varepsilon)$ contains a number τ with the property

$$\|f(t + \tau) - f(t)\| < \frac{\varepsilon}{2}$$

for all $t \in \mathbb{R}$.

Since $f[0, \ell(\varepsilon)]$ is compact in $\mathbb{X}$, choose a finite sequence $t_1, \ldots, t_n \in$ $[0, \ell(\varepsilon)]$ such that

$$f(t) \in \bigcup_{i=1}^{n} B\left(f(t_i), \frac{\varepsilon}{2}\right).$$

Now let $t \in \mathbb{R}$, $\tau = \tau(t)$ such that $0 < t + \tau < \ell(\varepsilon)$, t_j an element of the sequence $t_1, \ldots, t_n$ such that $f(t + \tau) \in B(f(t_i), \frac{\varepsilon}{2})$. Then

$$\|f(t) - f(t_j)\| \le \|f(t) - f(t + \tau)\| + \|f(t + \tau) - f(t_j)\| < \varepsilon$$

so that

$$R(f) \subseteq \bigcup_{i=1}^{n} B\left(f(t_i), \frac{\varepsilon}{2}\right).$$

Since ε is arbitrary it follows that $R(f)$ is relatively compact.

Definition 4.5 (Bochner) A function $f \in BC(\mathbb{R}, \mathbb{X})$ is called (Bochner) almost periodic if for any sequence $(\sigma_n')_{n \in \mathbb{N}}$ of real numbers there exists a subsequence $(\sigma_n)_{n \in \mathbb{N}}$ of $(\sigma_n')_{n \in \mathbb{N}}$ such that $(f(t + \sigma_n))_{n \in \mathbb{N}}$ converges uniformly in $t \in \mathbb{R}$.

Theorem 4.6 ([41, 82]) *A function is Bohr almost periodic if and only if it is Bochner almost periodic.*

Proof Let $f : \mathbb{R} \mapsto \mathbb{X}$ be a Bohr almost periodic function and consider $\{f(t + s_n)\}$, the translates of f, for an arbitrary sequence $(s_n)_{n \in \mathbb{N}}$ of real numbers. Using the fact that the range of f is relatively compact (Theorem 4.4) and applying a diagonal procedure extraction, one can find a subsequence $\{f(t + t_n)\}$ of $\{f(t + s_n)\}$ which converges for each $t \in \mathbb{Q}$. Let us now show that $\{f(t + t_n)\}$ actually converges uniformly in $t \in \mathbb{R}$. For every $\varepsilon > 0$, let $\ell(\varepsilon)$ stand for the associated length (appearing in Definition 4.3) and choose $\delta(\varepsilon)$ (from the uniform continuity of f, see Definition 1.15). Next, subdivide the interval $[0, \ell(\varepsilon)]$ into d subdivisions denoted I_j for $j = 1, 2, \ldots, d$ whose length should not be greater than $\delta(\varepsilon)$. Further, in each I_k, we choose an $r_k \in I_k \cap \mathbb{Q}$. Let $n = N_0(\varepsilon)$ be chosen such that

$$\|f(r_k + t_n) - f(r_k + t_m)\| < \varepsilon \tag{4.2}$$

for all $n, m \ge N_0(\varepsilon)$ and $k = 1, 2, \ldots, d$.

Clearly, for every $t_0 \in \mathbb{R}$, there exists an ε-period denoted $\tau = \tau_0$ such that $0 \le t_0 + \tau \le \ell(\varepsilon)$ which yields $-t_0 \le \tau \le -t_0 + \ell(\varepsilon)$. Suppose that $t_1 = t_0 + \tau$ belongs to I_{k_0} and that $r_{k_0} \in I_{k_0}$ (r_{k_0} being the rational point chosen above).

Now

$$\|f(t_1 + t_n) - f(r_{k_0} + t_n)\| < \varepsilon \text{ and } \|f(t_1 + t_m) - f(r_{k_0} + t_m)\| < \varepsilon. \tag{4.3}$$

In view of Eqs. (4.2) and (4.3) it follows that,

$$\|f(t_0 + t_n) - f(t_0 + t_m)\|$$

$$\leq \|f(t_0 + t_n) - f(t_1 + t_n)\| + \|f(t_1 + t_n) - f(r_{k_0} + t_n)\|$$

$$+ \|f(r_{k_0} + t_n) - f(r_{k_0} + t_m)\| + \|f(r_{k_0} + t_m) - f(t_1 + t_m)\|$$

$$+ \|f(t_1 + t_m) - f(t_0 + t_m)\| < 5\varepsilon.$$

Since $t_0 \in \mathbb{R}$ is arbitrary, it follows that $\{f(t + t_n)\}$ converges uniformly in $t \in \mathbb{R}$ and so f is Bochner almost periodic.

Suppose that f is not Bohr almost periodic. Then there exists at least one $\varepsilon > 0$ such that $\ell(\varepsilon)$ does not exist. In particular, there exists an interval of length $\ell(\varepsilon)$ which has no ε-periods of f. Let $\sigma_1 \in \mathbb{R}$ be arbitrary and let $(\alpha_1, \beta_1) \subset \mathbb{R}$ be an interval of length larger than $2|\sigma_1|$ such that (α_1, β_1) has no ε-period for f. Setting $\sigma_2 := \frac{\alpha_1 + \beta_1}{2}$ one can see that $\sigma_2 - \sigma_1 \in (\alpha_1, \beta_1)$ which yields $\sigma_2 - \sigma_1$ cannot be an ε-period for f. Similarly, consider the interval $(\alpha_2, \beta_2) \subset \mathbb{R}$ of length larger than $2|\sigma_1| + 2|\sigma_2|$ such that (α_2, β_2) has no ε-period for f. Setting $\sigma_3 := \frac{\alpha_2 + \beta_2}{2}$ one can see that $\sigma_3 - \sigma_1$ and $\sigma_3 - \sigma_2$ belong to (α_2, β_2) which yields both $\sigma_3 - \sigma_1$ and $\sigma_3 - \sigma_2$ cannot be an ε-periods for f. Proceeding this way, one constructs a sequence $\sigma_1, \sigma_2, \sigma_3, \sigma_4, \ldots$ such that none of the differences $\sigma_i - \sigma_j$ for $i > j$ can be an ε-period for f. Consequently, for all $i > j$, we have $\sup \|f(t + \sigma_i) - f(t + \sigma_j)\| = \sup \|f(t + \sigma_i - \sigma_j) - f(t)\| \geq \varepsilon$ for all $t \in \mathbb{R}$, which means that f is not Bochner almost periodic. The proof is complete.

One of the consequences of Theorem 4.6 is that the space of almost periodic functions, as stated in Theorem 4.7, is a vector space.

Theorem 4.7 *If $f, g : \mathbb{R} \mapsto \mathbb{X}$ are almost periodic functions and if $\lambda \in \mathbb{F}$, then the following properties hold,*

i) $f + g \in AP(\mathbb{X})$.
ii) $\lambda f \in AP(\mathbb{X})$.
iii) *If g is $\mathbb{F}$-valued, then $fg \in AP(\mathbb{X})$.*
iv) *The function $t \mapsto f(t + \alpha)$ belongs to $AP(\mathbb{X})$ for all $\alpha \in \mathbb{R}$.*
v) *The function $t \mapsto f(\alpha t)$ belongs to $AP(\mathbb{X})$ for all $\alpha \in \mathbb{R}$.*

Proof The proof makes extensive use of the Bochner's almost periodicity as it is, according to Theorem 4.6, equivalent of the Bohr's almost periodicity.

i) Using the fact $f \in AP(\mathbb{X})$, for any sequence $(s_n'')_{n \in \mathbb{N}}$ of real numbers there exists a subsequence $(s_n')_{n \in \mathbb{N}}$ of $(s_n'')_{n \in \mathbb{N}}$ such that the sequence of functions $(f(t + s_n'))_{n \in \mathbb{N}}$ converges uniformly in $t \in \mathbb{R}$. Similarly, using the fact $g \in AP(\mathbb{X})$ it follows that there exists a subsequence $(s_n)_{n \in \mathbb{N}}$ of $(s_n')_{n \in \mathbb{N}}$ such that the sequence of functions $(g(t + s_n))_{n \in \mathbb{N}}$ converges uniformly in $t \in \mathbb{R}$. Now $(f(t + s_n))_{n \in \mathbb{N}}$ converges uniformly in $t \in \mathbb{R}$ as $(s_n)_{n \in \mathbb{N}}$ is also a subsequence of $(s_n'')_{n \in \mathbb{N}}$ and therefore $(f(t + s_n) + g(t + s_n))_{n \in \mathbb{N}}$ converges uniformly in $t \in \mathbb{R}$, that is, $f + g \in AP(\mathbb{X})$.

ii) From $f \in AP(\mathbb{X})$, for any sequence $(s'_n)_{n \in \mathbb{N}}$ of real numbers there exists a subsequence $(s_n)_{n \in \mathbb{N}}$ of $('_n)_{n \in \mathbb{N}}$ such that the sequence of functions $(f(t + s_n))_{n \in \mathbb{N}}$ converges uniformly in $t \in \mathbb{R}$. Now $(\lambda f(t + s_n))_{n \in \mathbb{N}}$ converges uniformly in $t \in \mathbb{R}$ and hence $\lambda f \in AP(\mathbb{X})$.

iii) The proof can be done as in i). Indeed, from $f \in AP(\mathbb{X})$, for any sequence $(s''_n)_{n \in \mathbb{N}}$ of real numbers there exists a subsequence $(s'_n)_{n \in \mathbb{N}}$ of $(s''_n)_{n \in \mathbb{N}}$ such that the sequence of functions $(f(t + s'_n))_{n \in \mathbb{N}}$ converges uniformly in $t \in \mathbb{R}$. Since $g \in AP(\mathbb{F})$ it follows that there exists a subsequence $(s_n)_{n \in \mathbb{N}}$ of $(s'_n)_{n \in \mathbb{N}}$ such that the sequence of functions $(g(t + s_n))_{n \in \mathbb{N}}$ converges uniformly in $t \in \mathbb{R}$. Now $(f(t + s_n))_{n \in \mathbb{N}}$ converges uniformly in $t \in \mathbb{R}$ as $(s_n)_{n \in \mathbb{N}}$ is also a subsequence of $(s''_n)_{n \in \mathbb{N}}$ and therefore $(f(t + s_n).g(t + s_n)_{n \in \mathbb{N}}$ converges uniformly in $t \in \mathbb{R}$, that is, $fg \in AP(\mathbb{X})$.

iv) From $f \in AP(\mathbb{X})$, for any sequence $(s'_n)_{n \in \mathbb{N}}$ of real numbers there exists a subsequence $(s_n)_{n \in \mathbb{N}}$ of $(s'_n)_{n \in \mathbb{N}}$ such that the sequence of functions $(f(s + s_n))_{n \in \mathbb{N}}$ converges uniformly in $s \in \mathbb{R}$. In particular, letting $s = t + \alpha$ it follows that $(f(t + \alpha + s_n))_{n \in \mathbb{N}}$ converges uniformly in $t \in \mathbb{R}$ and hence $\theta \mapsto f(\theta + \alpha)$ belongs to $AP(\mathbb{X})$ for all $\alpha \in \mathbb{R}$.

v) From $f \in AP(\mathbb{X})$, for any sequence $(s'_n)_{n \in \mathbb{N}}$ of real numbers there exists a subsequence $(s_n)_{n \in \mathbb{N}}$ of $(s'_n)_{n \in \mathbb{N}}$ such that the sequence of functions $(f(s + s_n))_{n \in \mathbb{N}}$ converges uniformly in $s \in \mathbb{R}$. In particular, letting $s = \alpha t$ it follows that $(f(\alpha t + s_n))_{n \in \mathbb{N}}$ converges uniformly in $t \in \mathbb{R}$ and hence $\theta \mapsto f(\alpha \theta)$ belongs to $AP(\mathbb{X})$.

Theorem 4.8 *Let $\mathbb{X}$ be a uniformly convex Banach space. If $f : \mathbb{R} \mapsto \mathbb{X}$ is almost periodic, then the function defined by,*

$$F(t) = \int_{t_0}^{t} f(\sigma) d\sigma$$

is almost periodic if and only if $\sup_{t \in \mathbb{R}} \| F(t) \| < \infty$.

Proof We refer the reader to the book by Corduneanu [40, Proof of Theorem 6.20, pages 179–180].

4.2.2 Fourier Series Representation

Definition 4.9 If $f : \mathbb{R} \mapsto \mathbb{X}$ is a continuous function and if the limit

$$\lim_{r \to \infty} \frac{1}{2r} \int_{-r}^{r} f(t) dt$$

exists, we then call it the *mean value* of the function f and denote it $M(f)$.

Theorem 4.10 *If* $f : \mathbb{R} \mapsto \mathbb{X}$ *is almost periodic, then the mean value of* f,

$$M(f) := \lim_{r \to \infty} \frac{1}{2r} \int_{-r}^{r} f(t)dt$$

exists. Furthermore,

$$\lim_{r \to \infty} \frac{1}{2r} \int_{-r+a}^{r+a} f(t)dt = M(f)$$

exists uniformly in $a \in \mathbb{R}$.

Proof We refer the reader to the book by Levitan and Zhikov [82, Proof of Property 1—The Mean Value Theorem, pages 22–23].

Recall that the mean $M(f)$ of an almost periodic function f can be written in various way. Among others, we have

$$M(f) = \lim_{r \to \infty} \frac{1}{r} \int_{0}^{r} f(t)dt = \lim_{r \to \infty} \frac{1}{2r} \int_{-r}^{r} f(t)dt = \lim_{r \to \infty} \frac{1}{r} \int_{-\infty}^{0} f(t)dt.$$

Corollary 4.11 ([41]) *If* $f : \mathbb{R} \mapsto \mathbb{X}$ *is a continuous* ω-*periodic function (* $f(t + \omega) = f(t)$ *for all* $t \in \mathbb{R}$), *then its mean value exists and is given by*

$$M(f) = \frac{1}{\omega} \int_{0}^{\omega} f(\sigma)d\sigma.$$

Proposition 4.12 *Let* $f, g : \mathbb{R} \mapsto \mathbb{X}$ *be almost periodic functions and let* $\alpha \in \mathbb{C}$. *Then the following properties hold,*

i) *If* f *is* $\mathbb{C}$-*valued, then* $M(\overline{f}) = \overline{M(f)}$.
ii) $M(\alpha f) = \alpha M(f)$.
iii) *If* f *is* $\mathbb{R}$-*valued, then* $M(f) \geq 0$ *for all* $f \geq 0$.
iv) $M(f + g) = M(f) + M(g)$.
v) *If* $(f_n)_{n \in \mathbb{N}}$ *is a sequence of almost periodic functions, which converges uniformly to* f, *then*

$$\lim_{n \to \infty} M(f_n) = M(f).$$

Proof The proof is left to the reader as an exercise.

Clearly, if $f \in AP(\mathbb{X})$ and if $\lambda \in \mathbb{R}$, then the function defined by $t \mapsto f(t)e^{-i\lambda t}$ belongs to $AP(\mathbb{X})$. Define the *Fourier coefficients* of f by setting

$$a(\lambda, f) := M(\{f(t)e^{-i\lambda t}\}) = \lim_{r \to \infty} \frac{1}{2r} \int_{-r}^{r} f(t)e^{-i\lambda t}dt.$$

Definition 4.13 If $f \in AP(\mathbb{X})$, then the numbers $\lambda_1, \lambda_2, \ldots, \lambda_n, \ldots$ for which $a(\lambda_k, f) \neq 0$ are called the Fourier exponents of f. The set of Fourier exponents is called the Bohr spectrum and is denoted by $\sigma_b(f)$.

If $f \in AP(\mathbb{X})$, then $a(\lambda, f)$ is nonzero at most at countably many points.

Definition 4.14 If $f \in AP(\mathbb{X})$ and if $\sigma_b(f) = \{\lambda_k : k = 1, 2, \ldots\}$, then the series

$$\widehat{f}(t) := \sum_{k=1}^{\infty} a(\lambda_k, f) e^{i\lambda_k t}$$

will be called the Fourier series associated with f.

Theorem 4.15 ([82, 83]) *If $f \in AP(\mathbb{X})$, then for every $\varepsilon > 0$ there exists a trigonometric polynomial*

$$P_\varepsilon(t) = \sum_{k=1}^{n} a_k e^{i\lambda_k t}$$

where $a_k \in \mathbb{X}$ and $\lambda_k \in \sigma_b(f)$ such that $\|f(t) - P_\varepsilon(t)\| < \varepsilon$ for all $t \in \mathbb{R}$.

Proposition 4.16 ([41, 47]) *If $f, g \in AP(\mathbb{X})$ and if $\widehat{f} = \widehat{g}$, then $f = g$.*

For more on Fourier series theory for almost periodic functions and related issues we refer to Besicovitch [26] and Corduneanu [40, 41].

4.2.3 Composition of Almost Periodic Functions

Definition 4.17 ([47]) A jointly continuous function $F : (t, x) \mapsto F(t, x)$ is called almost periodic if $t \mapsto F(t, x)$ is almost periodic uniformly in $x \in B$, where B is any bounded subset of $\mathbb{X}$. That is, for each $\varepsilon > 0$ there exists $\ell(\varepsilon) > 0$ such that every interval of length $\ell(\varepsilon) > 0$ contains a number τ with the property

$$\|F(t + \tau, x) - F(t, x)\| < \varepsilon$$

for all $t \in \mathbb{R}$ and $x \in B$.

The collection of such functions will be denoted $AP(\mathbb{R} \times \mathbb{X})$ or $AP(\mathbb{R} \times \mathbb{X}, \mathbb{X})$.

To study the existence of almost periodic solutions to semilinear differential equations, one makes use of the following composition theorems whose proofs are omitted.

Theorem 4.18 ([47]) *Let $F : \mathbb{R} \times \mathbb{X} \mapsto \mathbb{X}$, $(t, x) \mapsto F(t, x)$ be an almost periodic function in $t \in \mathbb{R}$ uniformly in $x \in B$ where $B \subset \mathbb{X}$ is an arbitrary bounded subset.*

Suppose that F is Lipschitz in $x \in \mathbb{X}$ uniformly in $t \in \mathbb{R}$, i.e., there exists $L > 0$ such that

$$\|F(t, x) - F(t, y)\| \leq L \|x - y\|, \quad \text{for all } x, y \in \mathbb{X}, \ t \in \mathbb{R}. \qquad (4.4)$$

If $g : \mathbb{R} \mapsto \mathbb{X}$ is almost periodic, then the function $\Gamma(t) = F(t, g(t)) : \mathbb{R} \mapsto \mathbb{X}$ is also almost periodic.

Theorem 4.19 ([47]) *Let $F : \mathbb{R} \times \mathbb{X} \mapsto \mathbb{X}$ be an almost periodic function. Suppose $u \mapsto F(t, u)$ is uniformly continuous on every bounded subset $B' \subset \mathbb{X}$ uniformly for $t \in \mathbb{R}$. If $g \in AP(\mathbb{X})$, then $\Gamma : \mathbb{R} \to \mathbb{X}$ defined by $\Gamma(\cdot) := F(\cdot, g(\cdot))$ belongs to $AP(\mathbb{X})$.*

Theorem 4.20 ([47]) *If $F : \mathbb{R} \times \mathbb{X} \mapsto \mathbb{X}$, $(t, x) \mapsto F(t, x)$ is a jointly continuous function such that $t \mapsto F(t, x)$ is almost periodic uniformly in $x \in K$ where $K \subset \mathbb{X}$ is an arbitrary compact subset and if $g \in AP(\mathbb{X})$, then $\Gamma : \mathbb{R} \to \mathbb{X}$ defined by $\Gamma(\cdot) := F(\cdot, g(\cdot))$ belongs to $AP(\mathbb{X})$.*

4.3   Asymptotically Almost Periodic Functions

Let $C_0(\mathbb{R}_+, \mathbb{X})$ stand for the collection of all continuous functions $\varphi : \mathbb{R}_+ \mapsto \mathbb{X}$ such that

$$\lim_{t \to \infty} \|\varphi(t)\| = 0.$$

4.3.1   Basic Definitions and Properties

Definition 4.21 A continuous function $f : \mathbb{R}_+ \mapsto \mathbb{X}$ is said to be asymptotically almost periodic if there exist $h \in AP(\mathbb{X})$ and $\varphi \in C_0(\mathbb{R}_+, \mathbb{X})$ such that

$$f(t) = h(t) + \varphi(t), \quad t \in \mathbb{R}_+.$$

The collection of all asymptotically almost periodic functions will be denoted $AAP(\mathbb{X})$.

Proposition 4.22 *A continuous function $f : \mathbb{R}_+ \mapsto \mathbb{X}$ is asymptotically almost periodic if and only if for any sequence $(\tau_n)_{n \in \mathbb{N}}$ with $\tau_n \to \infty$ as $n \to \infty$ there exists a subsequence $(\tau_{n_k})_{k \in \mathbb{N}}$ for which $f(t + \tau_{n_k})$ converges uniformly in $t \in \mathbb{R}_+$.*

Proof The proof is left to the reader as an exercise.

Lemma 4.23 ([47]) *If $f \in AP(\mathbb{X})$ such that $\lim_{t \to \infty} \|f(t)\| = 0$, then $f = 0$.*

Lemma 4.24 ([47]) *If $f \in AAP(\mathbb{X})$, that is, $f = h + \varphi$ where $h \in AP(\mathbb{X})$ and $\varphi \in C_0(\mathbb{R}_+, \mathbb{X})$, then*

$$\overline{\{h(t) : t \in \mathbb{R}\}} \subset \overline{\{f(t) : t \in \mathbb{R}_+\}}.$$

Proposition 4.25 ([47]) *The decomposition of asymptotically almost periodic functions is unique, that is, $AAP(\mathbb{X}) = AP(\mathbb{X}) \oplus C_0(\mathbb{R}_+, \mathbb{X})$.*

Proposition 4.26 ([47]) *If $f, g \in AP(\mathbb{X})$ and if there exists $t_0 \in \mathbb{R}_+$ such that $f(t) = g(t)$ for all $t \geq t_0$, then $f = g$.*

Proposition 4.27 ([47]) *If $(f_n)_{n \in \mathbb{N}} \subset AAP(\mathbb{X})$ such that f_n converges uniformly to f on $\mathbb{R}_+$, that is,*

$$\lim_{n \to \infty} \|f_n - f\|_{AAP} = 0.$$

Then $f \in AAP(\mathbb{X})$.

4.3.2 Composition of Asymptotically Almost Periodic Functions

Let $C_0(\mathbb{R}_+ \times \mathbb{X})$ stand for the collection of all jointly continuous functions $\Psi : \mathbb{R}_+ \times \mathbb{X} \mapsto \mathbb{X}$ such that

$$\lim_{t \to \infty} \|\Psi(t, x)\| = 0$$

uniformly in $x \in K$ where $K \subset \mathbb{X}$ is an arbitrary bounded subset.

Definition 4.28 A jointly continuous function $F : \mathbb{R}_+ \times \mathbb{X} \mapsto \mathbb{X}$, $(t, x) \mapsto F(t, x)$ is called asymptotically almost periodic if $t \mapsto F(t, x)$ is asymptotically almost periodic uniformly in $x \in K$, where K is an arbitrary bounded subset of $\mathbb{X}$. That is, $F = H + \Phi$ where $H \in AP(\mathbb{R} \times \mathbb{X})$ and $\Phi \in C_0(\mathbb{R}_+ \times \mathbb{X})$. The collection of all such asymptotically almost periodic functions will be denoted $AAP(\mathbb{R}_+ \times \mathbb{X})$ or $AAP(\mathbb{R}_+ \times \mathbb{X}, \mathbb{X})$.

Theorem 4.29 ([47]) *Let $F : \mathbb{R}_+ \times \mathbb{X} \mapsto \mathbb{X}$, $(t, x) \mapsto F(t, x)$ be an asymptotically almost periodic function in $t \in \mathbb{R}_+$ uniformly in $x \in B$, where B is arbitrary bounded subset. Letting $F = H + \Phi$ where $H \in AP(\mathbb{R} \times \mathbb{X})$ and $\Phi \in C_0(\mathbb{R}_+ \times \mathbb{X})$, we suppose that both H and Φ are Lipschitz in $x \in \mathbb{X}$ uniformly in t, i.e., there exists $L_1, L_2 > 0$ such that*

$$\|H(t, x) - H(t, y)\| \leq L_1 \|x - y\| \tag{4.5}$$

for all $x, y \in \mathbb{X}$ *and* $t \in \mathbb{R}$, *and*

$$\|\Phi(t, x) - \Phi(t, y)\| \leq L_2 \|x - y\| \tag{4.6}$$

for all $x, y \in \mathbb{X}$ *and* $t \in \mathbb{R}_+$

If $g : \mathbb{R}_+ \mapsto \mathbb{X}$ *is asymptotically almost periodic, then the function* $\Gamma(t) = F(t, g(t)) : \mathbb{R}_+ \mapsto \mathbb{X}$ *is also asymptotically almost periodic.*

Theorem 4.30 ([47]) *Let* $F : \mathbb{R}_+ \times \mathbb{X} \mapsto \mathbb{X}$, $(t, x) \mapsto F(t, x)$ *be asymptotically almost periodic in* $t \in \mathbb{R}_+$ *uniformly in* $x \in B$ *where* B *is arbitrary bounded subset. Letting* $F = H + \Phi$ *where* $H \in AP(\mathbb{R} \times \mathbb{X})$ *and* $\Phi \in C_0(\mathbb{R}_+ \times \mathbb{X})$, *we suppose that* $u \mapsto H(t, u)$ *is uniformly continuous on every bounded subset* $B \subset \mathbb{X}$ *uniformly for* $t \in \mathbb{R}$ *and that* Φ *is Lipschitz in* $x \in \mathbb{X}$ *uniformly in* $t \in \mathbb{R}_+$, *i.e., there exists* $L > 0$ *such that*

$$\|\Phi(t, x) - \Phi(t, y)\| \leq L \|x - y\|, \quad \text{for all } x, y \in \mathbb{X}. \ t \in \mathbb{R}_+. \tag{4.7}$$

If $g : \mathbb{R}_+ \mapsto \mathbb{X}$ *is asymptotically almost periodic, then the function* $\Gamma(t) = F(t, g(t)) : \mathbb{R}_+ \mapsto \mathbb{X}$ *is also asymptotically almost periodic.*

Theorem 4.31 ([47]) *Let* $F : \mathbb{R}_+ \times \mathbb{X} \mapsto \mathbb{X}$ *be an asymptotically almost periodic function. Suppose* $u \mapsto F(t, u)$ *is uniformly continuous on every bounded subset* $B \subset \mathbb{X}$ *uniformly for* $t \in \mathbb{R}_+$. *If* $g \in AAP(\mathbb{X})$, *then* $\Gamma : \mathbb{R}_+ \to \mathbb{X}$ *defined by* $\Gamma(\cdot) := F(\cdot, g(\cdot))$ *belongs to* $AAP(\mathbb{X})$.

Theorem 4.32 ([47]) *If* $F : \mathbb{R}_+ \times \mathbb{X} \mapsto \mathbb{X}$, $(t, x) \mapsto F(t, x)$ *such that* $t \mapsto F(t, x)$ *is asymptotically almost periodic uniformly in* $x \in K$ *where* $K \subset \mathbb{X}$ *is an arbitrary compact subset and if* $g \in AAP(\mathbb{X})$, *then* $\Gamma : \mathbb{R} \to \mathbb{X}$ *defined by* $\Gamma(\cdot) := F(\cdot, g(\cdot))$ *belongs to* $AAP(\mathbb{X})$.

4.4 Almost Periodic Sequences

4.4.1 Basic Definitions

Let $I = \mathbb{Z}_+$ or $\mathbb{Z}$. Recall that $\ell^\infty(I)$ stands for the Banach space of all bounded $\mathbb{X}$-valued sequences equipped with the *sup-norm* defined for each $x = \{x(t)\}_{t \in I} \in \ell^\infty(I)$, by

$$\|x\|_\infty = \sup_{t \in I} \|x(t)\|.$$

Define

$$N(\mathbb{Z}_+) := \left\{ x = (x(t))_{t \in \mathbb{Z}_+} \in \ell^\infty(\mathbb{Z}_+) : \lim_{t \to \infty} x(t) = 0 \right\}.$$

Definition 4.33 An $\mathbb{X}$-valued sequence $x = \{x(t)\}_{t \in \mathbb{Z}}$ is called (Bohr) almost periodic if for each $\varepsilon > 0$, there exists a positive integer $N_0(\varepsilon)$ such that any set consisting of $N_0(\varepsilon)$ consecutive integers contains at least one integer τ with the property

$$\|x(t + \tau) - x(t)\| < \varepsilon, \ \forall t \in \mathbb{Z}.$$

As in the continuous setting, the integer τ is called an ε-period (or an ε-translation number) of the sequence $x = \{x(t)\}_{t \in \mathbb{Z}}$. The collection of all the ε-periods for the sequence x is denoted by $\mathscr{P}(\varepsilon, x)$. The collection of all almost periodic $\mathbb{X}$-valued sequences on $\mathbb{Z}$ will be denoted by $AP(\mathbb{Z})$.

Definition 4.34 An $\mathbb{X}$-valued sequence $x = \{x(t)\}_{t \in \mathbb{Z}}$ is called (Bochner) almost periodic if for every sequence $\{h(t)\}_{t \in \mathbb{Z}} \subset \mathbb{Z}$ there exists a subsequence $\{h(K_s)\}_{s \in \mathbb{Z}}$ such that $\{x(t + h(K_s))\}_{s \in \mathbb{Z}}$ converges uniformly in $t \in \mathbb{Z}$.

4.4.2   Properties of Almost Periodic Sequences

If $x = \{x(t)\}_{t \in \mathbb{Z}}$ and $y = \{y(t)\}_{t \in \mathbb{Z}}$ belong to $AP(\mathbb{Z})$ and if $\lambda, \mu \in \mathbb{F}$, then the following properties hold,

i) $\lambda x \in AP(\mathbb{Z})$;
ii) $\lambda x + \mu y \in AP(\mathbb{Z})$;
iii) If x is $\mathbb{F}$-valued, then $xy \in AP(\mathbb{Z})$ where $xy = \{x(t)y(t)\}_{t \in \mathbb{Z}}$;
iv) $\dfrac{x}{y} \in AP(\mathbb{Z})$ provided y is $\mathbb{F}$-valued and $|y(t)| \geq \alpha > 0$ for all $t \in \mathbb{Z}$; and
v) for any fixed $s \in \mathbb{Z}$, $x_s \in AP(\mathbb{X})$ where $x_s(t) = x(t + s)$ for all $t \in \mathbb{Z}$.

In view of the above, it follows that $AP(\mathbb{Z})$ is a vector space, which in fact is a Banach space when it is equipped with the sup norm $\| \cdot \|_\infty$.

Theorem 4.35 ([63]) *A necessary and sufficient condition for a sequence* $x = \{x(t)\}_{t \in \mathbb{Z}}$ *to be almost periodic is the existence of an almost periodic function* $f \in AP(\mathbb{X})$ *such that* $x(t) = f(t)$ *for all* $t \in \mathbb{Z}$.

Proof Let $f \in AP(\mathbb{X})$ and let $\{\alpha_j\}_j$ be a sequence of integers. Define, $x(t + \alpha_j) := f(t + \alpha_j)$ for $j \in \mathbb{N}$ and $t \in \mathbb{Z}$. Using the fact that the function f is (Bochner) almost periodic, it follows that there exists a subsequence $\{\beta_i\}_i$ of $\{\alpha_j\}_j$ such that $\{f(s + \beta_i)\}_i$ converges uniformly in $s \in \mathbb{Z}$. Consequently, $(x(t + \beta_j))_j$ converges uniformly in $t \in \mathbb{Z}$ which yields the almost periodicity of $(x(t))_{t \in \mathbb{Z}}$.

Conversely, suppose that $x = \{x(t)\}_{t \in \mathbb{Z}}$ is an almost periodic sequence. Define the map $f : \mathbb{R} \mapsto \mathbb{X}$ as follows:

$$f(t) = x(s) + (t - s)(x(s + 1) - x(s)), \ \ s \leq t < s + 1, \ \ \text{for all } s \in \mathbb{Z}.$$

Obviously, $f(t) = x(t)$ for all $t \in \mathbb{Z}$. Further, one can easily see that if $\tau \in \mathscr{P}(\frac{\varepsilon}{3}, x)$, then $\tau \in \mathscr{P}(\varepsilon, f)$.

Theorem 4.36 *Every almost periodic sequence is bounded.*

Proof Using the fact $x(t) = f(t)$ for all $t \in \mathbb{Z}$ (Theorem 4.35), for some almost periodic function $f : \mathbb{R} \mapsto \mathbb{X}$ (Theorem 4.35) it follows that

$$\sup_{t \in \mathbb{Z}} \|x(t)\| \leq \sup_{t \in \mathbb{R}} \|f(t)\| < \infty.$$

Consequently, the sequence $(x(t))_{t \in \mathbb{Z}}$ is bounded.

Proposition 4.37 *Let $x^m = \{x^m(t)\}_{t \in \mathbb{Z}}$ be an almost periodic sequence converging uniformly in $m \in \mathbb{Z}$ to x, then the sequence x is almost periodic.*

Proof The proof is left to the reader as an exercise.

Theorem 4.38 ([63]) *A sequence $x = \{x(t)\}_{t \in \mathbb{Z}}$ is Bochner almost periodic if and only if it is Bohr almost periodic.*

Proof Suppose that the sequence $x = \{x(t)\}_{t \in \mathbb{Z}}$ is Bochner almost periodic. If $x = \{x(t)\}_{t \in \mathbb{Z}}$ is not Bohr almost periodic, then there exists at least one $\varepsilon_0 > 0$ such that for any positive integer N_0, there exist N_0 consecutive integers which contain no ε_0-period for $\{x(t)\}_{t \in \mathbb{Z}}$. Let G_N denotes the group of N-consecutive integers and let σ_1 be an arbitrary integer. Choose σ_2 such that $\sigma_2 - \sigma_1 \in G_1$ and denote $G_1 := G_{v_1}$. Let v_2 be such that $v_2 > |\sigma_1 - \sigma_2|$ and choose σ_3 such that both $\sigma_3 - \sigma_1$ and $\sigma_3 - \sigma_2$ belong to G_{v_2}. Proceeding this way, one can construct a sequence

$$v_j \geq \max\{|\sigma_v - \sigma_\mu| : \ 1 \leq \mu \leq v \leq j\}$$

and choose σ_{j+1} such that $\sigma_{j+1} - \sigma_\mu \in G_{v_j}$ for $1 \leq \mu \leq j$. In fact, one may take $\sigma_{j+1} = \min\{g : g \in G_{v_j}\} + \max\{\sigma_\mu : 1 \leq \mu \leq j\}$.

Now

$$\sup_\sigma \|x(\sigma + \sigma_r) - x(\sigma + \sigma_s)\| = \sup_\sigma \|x(\sigma + \sigma_r - \sigma_s) - x(\sigma)\|$$

where $\sigma_r - \sigma_s \in G_{v_{r-1}}$ if $r \geq s$.

From the definition of G_{N_0} it follows that,

$$\sup_\sigma \|x(\sigma + \sigma_r) - x(\sigma + \sigma_s)\| \geq \varepsilon_0.$$

Let (σ_{j_i}) be a subsequence of the sequence $(\sigma_j)_j$ constructed previously such that $\{x(\sigma + \sigma_{j_i})\}$ converges uniformly in $\sigma \in \mathbb{Z}$. Consequently, there exists j_0 such that $r \geq j_0$ and $s \geq j_0$ yield

$$\|x(\sigma + \sigma_{j_r}) - x(\sigma + \sigma_{j_s})\| \geq \frac{\varepsilon_0}{2}$$

which is in contradiction with the property of the sequence $(\sigma_j)_j$ constructed above.

Conversely, suppose that the sequence $\{x(t)\}_{t\in\mathbb{Z}}$ is Bohr almost periodic and let $\{t_j\}_{j\in\mathbb{Z}}$ be a sequence of integers. For each $\varepsilon > 0$ there exists an integer $N_0 > 0$ such that between t_j and $t_j - N_0$ there exist an ε-period τ_j with $0 \le t_j - \tau_j \le N_0$. Setting $s_j = t_j - \tau_j$, one can see that s_j can only take a finite number (at most $N_0 + 1$) values. Consequently, there exists some s, $0 \le s \le N_0$ such that $s_j = s$ for infinite numbers of $j's$. Let these indexes be numbered as j_i, then we have

$$\left\|x(t+t_j) - x(t+s_j)\right\| = \left\|x(t+\tau_j+s_j) - x(t+s_j)\right\| < \varepsilon.$$

Hence,

$$\left\|x(t+t_j) - x(t+s_j)\right\| < \varepsilon$$

for all $t \in \mathbb{Z}$.

Now let $\{\varepsilon_r\}_r$ be a sequence such that $\varepsilon_r \to 0$ (for instance, $\varepsilon_r = \frac{1}{r}$). Now, from the sequence $\left\{x(t+t_j)\right\}_j$, consider a subsequence chosen so that

$$\left\|x(t+t_{j_i^1}) - x(n+s^1)\right\| \le \varepsilon_1.$$

Next, from the previous sequence, we take a new subsequence such that

$$\left\|x(n+t_{j_i^2}) - x(n+s^2)\right\| \le \varepsilon_2.$$

Repeating this procedure and for each $r \in \mathbb{N}$ we obtain a subsequence $\left\{x(t+t_{j_i^r})\right\}_i$ such that

$$\left\|x(t+t_{j_i^r}) - x(t+s^r)\right\| \le \varepsilon_r.$$

Now, for the diagonal sequence, $\left\{x(t+t_{j_i})\right\}_i$, for each $\varepsilon > 0$ take $k(\varepsilon) \in \mathbb{N}$ such that $\varepsilon_{k(\varepsilon)} < \frac{\varepsilon}{2}$, where ε_r belongs to the previous sequence $\{\varepsilon_r\}_{r\in\mathbb{N}}$.

Using the fact that the sequences $\{t_{j_r}\}$ and $\{t_{j_s}\}$ are both subsequences of $\left\{t_{j_i^{k(\varepsilon)}}\right\}$, for $r \ge k(\varepsilon)$ we have

$$\begin{aligned}
\left\|x(t+t_{j_r}) - x(t+t_{j_s})\right\| &\le \left\|x(t+t_{j_r}) - x(t+s^k)\right\| \\
&\quad + \left\|x(t+s^k) - x(t+t_{j_s})\right\| \\
&\le \varepsilon_{k(\varepsilon)} + \varepsilon_{k(\varepsilon)} \\
&\le \varepsilon.
\end{aligned}$$

Therefore, the sequence $\left\{x(t+t_{j_i})\right\}_i$ is a Cauchy sequence and hence converges uniformly in $t \in \mathbb{Z}$.

4.4.3 Composition of Almost Periodic Sequences

Definition 4.39 A sequence $F : \mathbb{Z} \times \mathbb{X} \mapsto \mathbb{X}$, $(t, u) \mapsto F(t, u)$ is called almost periodic in $t \in \mathbb{Z}$ uniformly in $u \in B$ where $B \subset \mathbb{X}$ is an arbitrary bounded subset, if for each $\varepsilon > 0$ there exists a positive integer $N_0(\varepsilon)$ such that among any $N_0(\varepsilon)$ consecutive integers, there exists at least one integer s with the property:

$$\| F(t + s, u) - F(t, u) \| < \varepsilon$$

for all $t \in \mathbb{Z}$ and $u \in B$.

We have the following composition theorems (of almost periodic sequences) which are used to deal with the existence of almost periodic solutions to semilinear difference equations. Their proofs are omitted as there are particular cases of composition theorems for almost periodic functions.

Theorem 4.40 *Suppose that $F : \mathbb{Z} \times \mathbb{X} \to \mathbb{X}$, $(t, u) \mapsto F(t, u)$ is almost periodic in $t \in \mathbb{Z}$ uniformly in $u \in B$, where $B \subset \mathbb{X}$ is a bounded subset of $\mathbb{X}$. If in addition, F is Lipschitz in $u \in \mathbb{X}$ uniformly in $t \in \mathbb{Z}$ that is, there exists $L > 0$ such that*

$$\| F(t, u) - F(t, v) \| \leq L \| u - v \| \;\; \text{for all } u, v \in \mathbb{X}, \, t \in \mathbb{Z},$$

then for every $\mathbb{X}$-valued almost periodic sequence $x = \{x(t)\}_{t \in \mathbb{Z}}$, the $\mathbb{X}$-valued sequence $y(t) = F(t, x(t))$ is almost periodic.

Theorem 4.41 *If $F : \mathbb{Z} \times \mathbb{X} \mapsto \mathbb{X}$, $(t, x) \mapsto F(t, x)$ such that $t \mapsto F(t, x)$ is almost periodic uniformly in $x \in K$ where $K \subset \mathbb{X}$ is an arbitrary compact subset and if $t \mapsto z(t)$ is almost periodic, then the sequence $y : \mathbb{Z} \to \mathbb{X}$ defined by $y(\cdot) := F(\cdot, z(\cdot))$ is almost periodic.*

4.5 Asymptotically Almost Periodic Sequences

4.5.1 Basic Definitions

Definition 4.42 An $\mathbb{X}$-valued sequence $x = \{x(t)\}_{t \in \mathbb{Z}_+}$ is called asymptotically almost periodic if for each $\varepsilon > 0$, there exist two positive integers N, M such that any set consisting of $N + 1$ consecutive positive integers contains at least one integer τ with the property

$$\| x(t + \tau) - x(t) \| < \varepsilon, \; \forall t \geq M.$$

Definition 4.42 is equivalent to the following definition:

Definition 4.43 An $\mathbb{X}$-valued sequence $x = \{x(t)\}_{t \in \mathbb{Z}_+}$ is called asymptotically almost periodic if for each $\varepsilon > 0$, there exist two positive integers N, M such that, any set consisting of $N + 1$ consecutive integers contains at least one integer τ with the property

$$\|x(t + \tau) - x(t)\| < \varepsilon, \ \ t \geq N, \ \ t + \tau \geq M.$$

The integer τ is called an asymptotic ε-period of the sequence $x = \{x(t)\}_{t \in \mathbb{Z}_+}$. The collection of all asymptotically almost periodic $\mathbb{X}$-valued sequences defined above will be denoted $AAP(\mathbb{Z})$.

4.5.2 Properties of Asymptotically Almost Periodic Sequences

Theorem 4.44 *If the $\mathbb{X}$-valued sequence $x = \{x(t)\}_{t \in \mathbb{Z}_+}$ is asymptotically almost periodic, then it can be decomposed as*

$$x(t) = u(t) + \upsilon(t), \ \ t \in \mathbb{Z}_+,$$

where $u = \{u(t)\}_{t \in \mathbb{Z}} \in AP(\mathbb{Z})$ and $\{\upsilon(t)\}_{t \in \mathbb{Z}_+} \in N(\mathbb{Z}_+)$.

Proof We refer the reader to this article by Fan [56].

Lemma 4.45 *If $x = \{x(t)\}_{t \in \mathbb{Z}} \in AP(\mathbb{Z})$ and $\lim\limits_{t \to \infty} x(t) = 0$, then $x(t) = 0$ for all $t \in \mathbb{Z}$.*

Proof We use the fact that there exists $f \in AP(\mathbb{X})$ such that $x(t) = f(t)$ for all $t \in \mathbb{Z}$. Moreover, it is well known that any almost periodic function h that goes to 0 as $x \to 0$ is identically equal to zero. Consequently, $f = 0$ which yields $x = 0$.

Lemma 4.46 *The decomposition of an asymptotically almost periodic sequence is unique. That is, $AP(\mathbb{Z}) \cap N(\mathbb{Z}_+) = \{0\}$.*

Proof Suppose that $x = \{x(t)\}_{t \in \mathbb{Z}_+}$ can be decomposed as $x(t) = u(t) + \upsilon(t)$ and $x(t) = v(t) + \delta(t)$, where $u = \{u(t)\}_{t \in \mathbb{Z}}$, $v = \{v(t)\}_{t \in \mathbb{Z}} \in AP(\mathbb{Z})$ and $\{\upsilon(t)\}_{t \in \mathbb{Z}_+}$, $\{\delta(t)\}_{t \in \mathbb{Z}_+} \in N(\mathbb{Z}_+)$. Clearly, $u(t) - v(t) = \delta(t) - \upsilon(t) \in AP(\mathbb{Z}) \cap N(\mathbb{Z}_+)$. By Lemma 4.45, we deduce that $u = v$ and $\upsilon = \delta$.

4.6 Exercises

1. If $f, g : \mathbb{R} \mapsto \mathbb{X}$ are almost periodic functions and if $\lambda \in \mathbb{F}$, then the following properties hold,

   i) If g is $\mathbb{F}$-valued such that $0 < M \leq |g(t)|$ for each $t \in \mathbb{R}$ for some constant M, then the quotient function $t \mapsto fg^{-1}(t)$ is almost periodic.

   ii) Let $(f_n)_{n\in\mathbb{N}}$ be a sequence of almost periodic functions such that there
      exists $h \in C(\mathbb{R}, \mathbb{X})$ with $\|f_n - h\|_\infty \to 0$ as $n \to \infty$. Then h is almost
      periodic.

  iii) Show that if $f \in AP(\mathbb{X})$ and $g \in L^1(\mathbb{R})$, then their convolution defined by

$$(f * g)(t) = \int_{-\infty}^{\infty} f(s)g(t-s)ds$$

   belongs to $AP(\mathbb{X})$.

2. Show that $(AP(\mathbb{X}), \|\cdot\|_\infty)$ is a Banach space.
3. Show that the function defined by $f(t) = \sum_{k=0}^{d} a_k e^{i\lambda_k t}$ for all $t \in \mathbb{R}$ is almost
   periodic.
4. Show that if $f : \mathbb{R} \mapsto \mathbb{X}$ is almost periodic, then the mean value of f,

$$M(f) := \lim_{r\to\infty} \frac{1}{2r} \int_{-r}^{r} f(t)dt$$

exists. Furthermore,

$$\lim_{r\to\infty} \frac{1}{2r} \int_{-r+a}^{r+a} f(t)dt = M(f)$$

exists uniformly in $a \in \mathbb{R}$.

5. If $f, g : \mathbb{R} \mapsto \mathbb{X}$ are almost periodic functions and if $\alpha, \beta \in \mathbb{C}$, show that

  a. If f is $\mathbb{C}$-valued, then $M(\overline{f}) = \overline{M(f)}$.
  b. $M(\alpha f + \beta g) = \alpha M(f) + \beta M(f)$.
  c. If f is $\mathbb{R}$-valued, then $M(f) \geq 0$ for all $f \geq 0$.
  d. Show that if f is $\mathbb{R}$-valued, $f \geq 0$, and $M(f) = 0$, then $f \geq 0$.
  e. If $(f_n)_{n\in\mathbb{N}}$ is a sequence of almost periodic functions, which converges
     uniformly to f, then

$$\lim_{n\to\infty} M(f_n) = M(f).$$

6. Show that $AP(\mathbb{R}) \cap L^p(\mathbb{R}) = \{0\}$ for $p \geq 1$.
7. Suppose $f \in AP(\mathbb{R})$. Is $f^N \in AP(\mathbb{R})$ (N being an positive integer)?
8. Show that a continuous function $f : \mathbb{R}_+ \mapsto \mathbb{X}$ is asymptotically almost periodic
   if and only if for any sequence $(\tau_n)_{n\in\mathbb{N}}$ with $\tau_n \to \infty$ as $n \to \infty$ there exists a
   subsequence $(\tau_{n_k})_{k\in\mathbb{N}}$ for which $f(t + \tau_{n_k})$ converges uniformly in $t \in \mathbb{R}_+$.
9. Prove Theorem 4.40.
10. Prove Theorem 4.41.

4.7 Comments

The materials on almost periodic and asymptotically almost periodic functions discussed in this chapter are mainly taken from the following sources: Besicovitch [26], Bohr [29], Corduneanu [40, 41], Diagana [45], Diagana [47], Bezandry and Diagana [27], Fink [58], Levitan and Zhikov [82], N'Guérékata [94, 95], and Zhang [114]. While the first part of the proof of Theorem 4.6 follows Levitan and Zhikov [82], the second one follows Corduneanu [41]. The proofs of Theorems 4.4 and 4.7 are taken from Diagana [47].

The part of the chapter on almost periodic and asymptotically almost periodic sequences is based upon Diagana [47], Diagana et al. [49], Halanay and Rasvan [63], and Fan [56]. The proofs of both Theorems 4.35 and 4.38 are taken from Halanay and Rasvan [63].

Details on the proof of Theorem 4.20 can be found in both Fink [58] and Zhang [114] for instance.

Chapter 5
Nonautonomous Difference Equations

5.1 Introduction

An autonomous difference equation is an equation of the form

$$x(t+1) = f_0(x(t)), \quad t \in \mathbb{Z}.$$

Although these equations play an important role when it comes to studying some models arising in population dynamics, they do not take into account some important parameters such as environmental fluctuations or seasonal changes. Nonautonomous difference equations, that is, equations of the form

$$x(t+1) = f_1(t, x(t)), \quad t \in \mathbb{Z},$$

seem to be more suitable to capture environmental fluctuations and seasonal changes, see, e.g., Elaydi [54].

The main objective of this chapter is two-fold. We first extend the theory of almost periodic sequences built in $\mathbb{Z}_+$ by Diagana et al. [49] to $\mathbb{Z}$. Next, we make extensive use of dichotomy techniques to find sufficient conditions for the existence of almost periodic solutions for the class of semilinear systems of difference equations given by

$$x(t+1) = A(t)x(t) + h(t, x(t)), \quad t \in \mathbb{Z} \tag{5.1}$$

where $A(t)$ is a $k \times k$ almost periodic matrix function defined on $\mathbb{Z}$, and the function $h : \mathbb{Z} \times \mathbb{R}^k \to \mathbb{R}^k$ is almost periodic in the first variable uniformly in the second one. As in the case of $\mathbb{Z}_+$, our existence results are, subsequently, applied to discretely reproducing populations with overlapping generations.

© Springer Nature Switzerland AG 2018

T. Diagana, *Semilinear Evolution Equations and Their Applications*,
https://doi.org/10.1007/978-3-030-00449-1_5

Recall once again that $\ell^\infty(\mathbb{Z})$, the Banach space of all bounded $\mathbb{R}^k$-valued sequences, is equipped with the *sup norm* defined for each $x = \{x(t)\}_{t \in \mathbb{Z}} \in \ell^\infty(\mathbb{Z})$, by

$$\|x\|_\infty = \sup_{t \in \mathbb{Z}} \|x(t)\|.$$

In order to deal with the existence of almost periodic solutions to the above-mentioned nonautonomous difference equations, we need to introduce the concepts of bi-almost periodicity and positively bi-almost periodicity for sequences.

Definition 5.1 A sequence $L : \mathbb{Z} \times \mathbb{Z} \mapsto \mathbb{R}^k$ is called bi-almost periodic if for every $\varepsilon > 0$, there exists a positive integer $N_0(\varepsilon)$ such that any set consisting of $N_0(\varepsilon)$ consecutive integers contains at least one integer σ for which

$$\|L(t + \sigma, s + \sigma) - L(t, s)\| < \varepsilon$$

for all $t, s \in \mathbb{Z}$. The collection of such sequences is denoted $bAP(\mathbb{Z} \times \mathbb{Z}, \mathbb{R}^k)$.

Let $\widetilde{\mathbb{T}}$ be the set defined by

$$\widetilde{\mathbb{T}} := \Big\{(t, s) \in \mathbb{Z} \times \mathbb{Z} : t \geq s\Big\}.$$

Definition 5.2 A sequence $L : \widetilde{\mathbb{T}} \mapsto \mathbb{R}^k$ is called positively bi-almost periodic if for every $\varepsilon > 0$, there exists a positive integer $N_0(\varepsilon)$ such that any set consisting of $N_0(\varepsilon)$ consecutive integers contains at least one integer σ for which

$$\|L(t + \sigma, s + \sigma) - L(t, s)\| < \varepsilon$$

for all $(t, s) \in \widetilde{\mathbb{T}}$. The collection of such sequences will be denoted $bAP(\widetilde{\mathbb{T}}, \mathbb{X})$.

Obviously, every bi-almost periodic sequence is positively bi-almost periodic with the converse being untrue.

Example 5.3 Classical examples of bi-almost periodic sequences L include those which are of the form $L(t, s) = h(t - s)$ for all $(t, s) \in \mathbb{Z} \times \mathbb{Z}$, where $h = (h(t)_{t \in \mathbb{Z}}$ is periodic, that is, there exists $0 \neq \omega \in \mathbb{Z}$ such that $h(t + \omega) = h(t)$ for all $t \in \mathbb{Z}$.

In this chapter, we are aimed at finding sufficient conditions for the existence of almost periodic solutions to the class of semilinear systems of difference equations given by

$$x(t + 1) = A(t)x(t) + f(t, x(t)), \quad t \in \mathbb{Z} \tag{5.2}$$

where $A(t)$ is a $k \times k$ almost periodic matrix function defined on $\mathbb{Z}$, and the function $f : \mathbb{Z} \times \mathbb{R}^k \to \mathbb{R}^k$ is almost periodic in the first variable uniformly in the second one.

To study the existence of solutions to Eq. (5.2), we make extensive use of the fundamental solutions to the system

$$x(t+1) = A(t)x(t), \quad t \in \mathbb{Z} \tag{5.3}$$

to examine almost periodic solutions of the system of difference equations

$$x(t+1) = A(t)x(t) + g(t), \quad t \in \mathbb{Z} \tag{5.4}$$

where $g : \mathbb{Z} \mapsto \mathbb{R}^k$ is almost periodic.

5.2   Discrete Exponential Dichotomy

Define the state transition matrix associated with $A(t)$ as follows

$$X(t,s) = \prod_{r=s}^{t-1} A(r), \quad X(t,t) = I,$$

for $t > s$.

Definition 5.4 ([65, Definition 7.6.4, p. 229]) Equation (5.3) is said to have a discrete exponential dichotomy if there exist $k \times k$ projection matrices $P(t)$ with $t \in \mathbb{Z}$ and positive constants M and $\beta \in (0, 1)$ such that,

  (i) $A(t)P(t) = P(t+1)A(t)$;
 (ii) The matrix $A(t)\Big(R(P(t))\Big)$ is an isomorphism from $R(P(t))$ onto $R(P(t+1))$;
(iii) $\|X(t,r)P(r)x\| \le M\beta^{r-t} \|x\|$, for $t < r, x \in \mathbb{R}^k$;
(iv) $\|X(t,r)(I - P(r))x\| \le M\beta^{t-r} \|x\|$, for $r \le t, x \in \mathbb{R}^k$.

By repeated application of [(i), Definition 5.4], we obtain

$$P(t)X(t,s) = X(t,s)P(s). \tag{5.5}$$

If Eq. (5.3) has a discrete dichotomy, then we define its associated Green function G by setting

$$G(t,s) = \begin{cases} -X(t,s)P(s) & \text{if } t < s, \\ X(t,s)(I - P(s)) & \text{if } t \ge s. \end{cases}$$

In view of the above, we have

$$\|G(t,s)\| \le \begin{cases} M\beta^{s-t} & \text{if } t < s, \\ M\beta^{t-s} & \text{if } t \ge s. \end{cases}$$

Remark 5.5 It should be mentioned that if $t \mapsto A(t)$ is almost periodic and if Eq. (5.3) has discrete dichotomy, then the Green operator function $G(t, s)Y \in bAP(\mathbb{T}, \mathbb{R}^k)$ uniformly for all Y in any bounded subset of $\mathbb{R}^k$.

We have the following characterization for the discrete exponential dichotomies:

Theorem 5.6 ([65, Theorem 7.6.5, p. 230]) *The following statements are equivalent,*

  i) *Equation (5.3) has a discrete exponential dichotomy;*
 ii) *For every bounded $\mathbb{R}^k$-valued sequence g, Eq. (5.4) has a unique bounded solution.*

If Eq. (5.3) has a discrete exponential dichotomy, then Theorem 5.6 ensures the existence and uniqueness of a bounded solution to Eq. (5.4) whenever $g : \mathbb{Z} \mapsto \mathbb{R}^k$ is a bounded sequence. Moreover, it can be shown that such a solution is given by

$$\overline{x}(t) = \sum_{r=-\infty}^{\infty} G(t, r+1)g(r)$$

$$= \sum_{r=-\infty}^{t-1} X(t, r+1)(I - P(r+1))g(r) - \sum_{r=t}^{\infty} X(t, r+1)P(r+1)g(r)$$

for all $t \in \mathbb{Z}$.

Theorem 5.7 *Suppose $t \mapsto A(t)$ is almost periodic and that Eq. (5.3) has a discrete exponential dichotomy. If $g \in AP(\mathbb{Z})$, then Eq. (5.4) has a unique almost periodic solution which can be expressed as*

$$\overline{x}(t) = \sum_{r=-\infty}^{t-1} X(t, r+1)(I - P(r+1))g(r) - \sum_{r=t}^{\infty} X(t, r+1)P(r+1)g(r). \quad (5.6)$$

Proof Since every almost periodic sequence is bounded, it follows from Theorem 5.6 that Eq. (5.4) has a unique bounded solution given by Eq. (5.6). Moreover,

$$\|\overline{x}(t)\| \le \left\{ \sum_{r=-\infty}^{t-1} \|X(t, r+1)(I - P(r+1))\| \right.$$

$$\left. + \sum_{r=t}^{\infty} \|X(t, r+1)P(r+1)\| \right\} \|g\|_\infty$$

$$\le \left\{ \frac{M}{1-\beta} + \frac{M\beta}{1-\beta} \right\} \|g\|_\infty$$

$$= \frac{M(1+\beta)}{1-\beta} \|g\|_\infty$$

which yields

$$\|\bar{x}\|_\infty \le \frac{M(1+\beta)}{1-\beta}\|g\|_\infty.$$

To complete the proof, one has to show that $\bar{x} \in AP(\mathbb{Z})$. For that, write $\bar{x} = M(g) - N(g)$ where

$$M(g)(t) := \sum_{r=-\infty}^{t-1} X(t, r+1)(I - P(r+1))g(r)$$

and

$$N(g)(t) = \sum_{r=t}^{\infty} X(t, r+1)P(r+1)g(r).$$

Let us show that $t \mapsto Mg(t)$ is almost periodic. Indeed, since g is almost periodic, for every $\varepsilon > 0$ there exists a positive integer $N_0(\varepsilon)$ such that any set consisting of $N_0(\varepsilon)$ consecutive integers contains at least one integer τ for which

$$\|g(t+\tau) - g(t)\| < \varepsilon$$

for all $t \in \mathbb{Z}$.

Setting $Q(t) = I - P(t)$, we obtain,

$$M(g)(t+\tau) - M(g)(t)$$

$$= \sum_{r=-\infty}^{t+\tau-1} X(t+\tau, r+1)Q(r+1)g(r) - \sum_{r=-\infty}^{t-1} X(t, r+1)Q(r+1)g(r)$$

$$= \sum_{r=-\infty}^{t-1} X(t+\tau, r+1+\tau)Q(r+\tau+1)g(r+\tau)$$

$$\quad - \sum_{r=-\infty}^{t-1} X(t, r+1)Q(r+1)g(r)$$

$$= \sum_{r=-\infty}^{t-1} X(t+\tau, r+1+\tau)Q(r+1+\tau)\big[g(r+\tau) - g(r)\big]$$

$$\quad + \sum_{r=-\infty}^{t-1} \big[X(t+\tau, r+1+\tau)Q(r+1+\tau) - X(t, r+1)Q(r+1)\big]g(r).$$

Clearly,

$$\left\| \sum_{r=-\infty}^{t-1} X(t+\tau, r+1+\tau) Q(r+1+\tau) \Big[g(r+\tau) - g(r) \Big] \right\| < c_1(\beta, M)\varepsilon$$

From Remark 5.5 it follows that

$$\left\| \sum_{r=-\infty}^{t-1} \Big[X(t+\tau, r+1+\tau) Q(r+1+\tau) \right.$$
$$\left. -X(t, r+1) Q(r+1) \Big] g(r) \right\| < c_2(\beta, M)\varepsilon,$$

and hence

$$\| M(g)(t+\tau) - M(g)(t) \| < c_3(\beta, M)\varepsilon$$

for each $t \in \mathbb{Z}$.

Using similar ideas as the previous ones, one can easily see that $N(g) \in AP(\mathbb{Z})$. This completes the proof.

Suppose that there exists $L > 0$ such that

$$\| f(t, x) - f(t, y) \| \le L \| x - y \|$$

for all $t \in \mathbb{R}$ and $x, y \in \mathbb{R}^k$.

Theorem 5.8 *Suppose that $t \mapsto A(t)$ is almost periodic and that Eq. (5.3) has a discrete exponential dichotomy. Further, we assume that $(t, w) \mapsto f(t, w)$ is almost periodic in $t \in \mathbb{Z}$ uniformly in $w \in B$ where $B \subset \mathbb{R}^k$ is an arbitrary bounded subset. Then Eq. (5.2) has a unique almost periodic solution given by*

$$z(t) = \sum_{r=-\infty}^{t-1} X(t, r+1) Q(r+1) f(r, z(r)) - \sum_{r=t}^{\infty} X(t, r+1) P(r+1) f(r, z(r)),$$
$$(5.7)$$

whenever L is small enough.

Proof Using the composition of almost periodic sequences (Theorem 4.40) it follows that $r \mapsto g(r) := f(r, z(r))$ belongs to $AP(\mathbb{Z})$ whenever $z \in AP(\mathbb{Z})$.

Let Δ be the nonlinear operator defined by

$$(\Delta z)(t) := \sum_{r=-\infty}^{\infty} G(t, r+1) g(r) \text{ for all } t \in \mathbb{Z}.$$

Using the proof of Theorem 5.7, one can easily see that Δ is well defined as it maps $AP(\mathbb{Z})$ into itself.

Now for all $u, v \in AP(\mathbb{Z})$,

$$\|(\Delta u)(t) - (\Delta v)(t)\| \leq \frac{M(1+\beta)}{1-\beta} \|f(t, u(t)) - f(t, u(t))\|,$$

and hence

$$\|\Delta u - \Delta v\|_\infty \leq \frac{ML(1+\beta)}{1-\beta} \|u - v\|_\infty.$$

Thus the nonlinear operator Δ is a strict contraction whenever L is small enough, that is,

$$\frac{ML(1+\beta)}{1-\beta} < 1.$$

To conclude, we make use of the classical Banach fixed point principle.

5.3   The Beverton-Holt Model with Overlapping Generations

To illustrate the results of the previous section, we consider the following theoretical discrete-time population model,

$$x(t+1) = f(t, x(t)) + \gamma x(t), \quad t \in \mathbb{Z} \tag{5.8}$$

where $x(t)$ is the total population size in generation t, $\gamma \in (0, 1)$ is the constant "probability" of surviving per generation, and $f : \mathbb{Z} \times \mathbb{R} \to \mathbb{R}$ models the birth or recruitment process.

In order to induce almost periodic effects on the population model, we consider the general model in the form,

$$x(t+1) = f(t, x(t)) + \gamma_t x(t), \quad t \in \mathbb{Z}. \tag{5.9}$$

where both $\{\gamma_t\}_{t \in \mathbb{Z}}$ and $f(t, x(t))$ belong to $AP(\mathbb{Z})$ and $\gamma_t \in (0, 1)$ for all $t \in \mathbb{Z}$.

Recall that Eq. (5.9) was studied by Franke and Yakubu [59] in $\mathbb{Z}_+$ and when recruitment function is of the form:

$$f(t, x(t)) = K_t(1 - \gamma_t), \tag{5.10}$$

and (with the periodic Beverton-Holt recruitment function)

$$f(t, x(t)) = \frac{(1-\gamma_t)\mu K_t x(t)}{(1-\gamma_t)K_t + (\mu - 1 + \gamma_t)x(t)}, \tag{5.11}$$

where the carrying capacity K_t is p-periodic, that is, $K_{t+p} = K_t$ for all $t \in \mathbb{Z}_+$ and $\mu > 1$ [43, 59].

Among other things, they have shown that the periodically forced recruitment functions Eqs. (5.10) and (5.11) generate globally attracting cycles in Eq. (5.9) (see details in [59]).

In this section, we extend these results to the almost periodic case in $\mathbb{Z}$. For that, we make use of Theorem 5.8 to show that if both $\{K_t\}_{t\in\mathbb{Z}}$ and $\{\gamma_t\}_{t\in\mathbb{Z}}$ are almost periodic, then Eq. (5.9) has a unique almost periodic solution.

Theorem 5.9 *Let*

$$f(t, x(t)) = \frac{(1 - \gamma_t)\mu K_t x(t)}{(1 - \gamma_t)K_t + (\mu - 1 + \gamma_t)x(t)},$$

where both $\{K_t\}_{t\in\mathbb{Z}}$ and $\{\gamma_t\}_{t\in\mathbb{Z}}$ are almost periodic, each $\gamma_t \in (0, 1)$, $K_t > 0$ and $\mu > 1$. Then Eq. (5.9) has a unique almost periodic solution whenever

$$\sup \{\gamma_t \mid_{t\in\mathbb{Z}}\} < \frac{1}{\mu + 1}.$$

Proof First of all, note that Eq. (5.9) is in the form of Eq. (5.2), where $A(t)$ and f can be taken respectively as follows

$$A(t) = \gamma_t,$$

and

$$f(t, x(t)) = \frac{(1 - \gamma_t)\mu K_t x(t)}{(1 - \gamma_t)K_t + (\mu - 1 + \gamma_t)x(t)}.$$

Now

$$|f(t, x) - f(t, y)|$$

$$\leq \frac{(1 - \gamma_t)^2 \mu K_t^2 |x - y|}{(1 - \gamma_t)^2 K_t^2 + (\mu - 1 + \gamma_t)(1 - \gamma_t)K_t(x + y) + (\mu - 1 + \gamma_t)^2 xy}$$

$$\leq \mu |x - y|.$$

Consequently, f is Lipschitz with the Lipschitz constant $L = \mu$. Similarly, take $M < \mu^{-1}$ and $\beta = \sup \{\gamma_t \mid_{t\in\mathbb{Z}}\}$. Clearly, Eq. (5.9) has a unique almost periodic solution whenever

$$\sup \{\gamma_t \mid_{t\in\mathbb{Z}}\} < \frac{1 - \mu M}{1 + \mu M}.$$

Similarly, if $f(t, x(t)) = K_t(1 - \gamma_t)$, then $f(t, x) - f(t, y) = 0$ which yields Eq. (5.9) has a unique almost periodic solution.

Corollary 5.10 *Let the recruitment function be* $f(t, x(t)) = K_t(1 - \gamma_t)$, *where both* $\{K_t\}_{t \in \mathbb{Z}}$ *and* $\{\gamma_t\}_{t \in \mathbb{Z}}$ *are almost periodic, each* $\gamma_t \in (0, 1)$ *and* $K_t > 0$. *Then Eq. (5.9) has a unique globally asymptotically stable almost periodic solution whenever*

$$\sup \{\gamma_t \mid_{t \in \mathbb{Z}}\} < 1.$$

5.4 Exercises

1. Prove Theorem 5.6.
2. Use dichotomy techniques to study the existence of almost periodic solutions to the semilinear difference equation with delay given by

$$u(t + 1) = A(t)u(t) + f(t, u(t), u(t - 1)), \quad t \in \mathbb{Z}$$

where $t \mapsto A(t)$ is a $d \times d$ almost periodic matrix and $f : \mathbb{Z} \times \mathbb{R}^d \times \mathbb{R}^d \to \mathbb{R}^d$ is almost periodic in $t \in \mathbb{Z}$ uniformly in the second and the third variables.

3. Use dichotomy techniques to study the existence of almost periodic solutions to the functional difference equation given by

$$u(t + 1) = A(t)u(t) + f(t, u(h_1(t)), u(h_2(t)), u(h_3(t))), \quad t \in \mathbb{Z}$$

where $t \mapsto A(t)$ is a $d \times d$ almost periodic matrix, the sequence $h_j : \mathbb{Z} \mapsto \mathbb{Z}$ with $h_j(\mathbb{Z}) = \mathbb{Z}$ for $j = 1, 2, 3$, and $f : \mathbb{Z} \times \mathbb{R}^d \times \mathbb{R}^d \times \mathbb{R}^d \to \mathbb{R}^d$ is almost periodic in $t \in \mathbb{Z}$ uniformly in the other variables.

5.5 Comments

The main references for this chapter include Diagana [47], Diagana et al. [49] and Henry [65]. Some parts of this chapter are based upon the following references: Diagana [46] and Araya et al. [17]. For additional readings on this topic, we refer to Diagana [47], Lizama and Mesquita [84], and Henry [65].

Chapter 6
Singular Difference Equations

6.1 Introduction

The mathematical problem which consists of studying the existence of solutions to singular difference equations with almost periodic coefficients is an important one as almost periodicity, according to Henson et al. [66], is more likely to accurately describe many phenomena occurring in population dynamics than periodicity. In the previous chapter, the existence of almost periodic solutions to some classes of nonautonomous non-singular difference equations was obtained. These results were utilized to study the effect of almost periodicity upon the Beverton-Holt model.

In this chapter, we study and establish the existence of Bohr (respectively, Besicovitch) almost periodic solutions to the following class of singular systems of difference equations,

$$Ax(t+1) + Bx(t) = f(t, x(t)) \tag{6.1}$$

where $f : \mathbb{Z} \times \mathbb{R}^N \to \mathbb{R}^N$ is Bohr (respectively, Besicovitch) almost periodic in $t \in \mathbb{Z}$ uniformly in the second variable, and A, B are $N \times N$ square matrices satisfying $\det A = \det B = 0$.

Recall that singular difference equations of the form Eq. (6.12) arise in many applications including optimal control, population dynamics, economics, and numerical analysis [52]. The main result discussed in this chapter can be summarized as follows: if $\lambda A + B$ is invertible for all $\lambda \in \mathbb{S}^1 = \{z \in \mathbb{C} : |z| = 1\}$ and if f is Bohr (respectively, Besicovitch) almost periodic in $t \in \mathbb{Z}$ uniformly in the second variable and under some additional conditions, then Eq. (6.12) has a unique Bohr (respectively, Besicovitch) almost periodic solution.

The chapter is organized as follows: Sect. 6.1 serves as an introduction but also provides preliminary tools needed in the sequel. In Sect. 6.2, some preliminary

© Springer Nature Switzerland AG 2018
T. Diagana, *Semilinear Evolution Equations and Their Applications*,
https://doi.org/10.1007/978-3-030-00449-1_6

results corresponding to the case $f(t, x(t)) = C(t)$ are obtained. Section 6.3 is devoted to the main results of this chapter. In Sect. 6.4, we make use of the main results in Sect. 6.3 to study the existence of Bohr (respectively, Besicovitch) almost periodic solutions for some second-order (and higher-order) singular systems of difference equations.

Let $x = (x(t))_{t \in \mathbb{Z}}$ be a sequence. Define $P(x)$ as follows

$$P(x) := \sup_{k \in \mathbb{Z}} \limsup_{N \to \infty} \left[\frac{1}{N} \sum_{j=k+1}^{k+N} \|x(j)\|^2 \right]^{\frac{1}{2}}.$$

Set

$$\widetilde{B} = \left\{ x = (x(t))_{t \in \mathbb{Z}} : P(x) < \infty \right\}.$$

It is not hard to see that P is a semi-norm on $\widetilde{B}$. Consider the following equivalence relation on $\widetilde{B}$: $x, y \in \widetilde{B}$, $x \sim y$ if and only if, $P(x - y) = 0$. The quotient space

$$B := \widetilde{B} / \sim$$

endowed with $P(\cdot)$ is a normed vector space.

Definition 6.1 A sequence $x = (x(t))_{t \in \mathbb{Z}}$ is called Besicovitch almost periodic if it belongs to the closure of trigonometric polynomials under the semi-norm P. The collection of all Besicovitch almost periodic sequences will be denoted $B^2(\mathbb{Z}, \mathbb{R}^N)$.

Definition 6.2 A sequence $F : \mathbb{Z} \times \mathbb{R}^N \mapsto \mathbb{R}^N$, $(t, u) \mapsto F(t, u)$ is called Besicovitch almost periodic in $t \in \mathbb{Z}$ if $t \mapsto F(t, u)$ belongs to $B^2(\mathbb{Z}, \mathbb{R}^N)$ uniformly in $u \in \mathbb{R}^N$.

6.2   The Case of a Linear Equation

In this section, we consider the case when the forcing term f does not depend on x, that is, $f(t, x(t)) = C(t)$ where $(C(t))_{t \in \mathbb{Z}}$ is almost periodic. Namely, we study the existence of almost periodic solutions for the singular difference equation

$$Ax(t + 1) + Bx(t) = C(t), \quad t \in \mathbb{Z} \tag{6.2}$$

where $C : \mathbb{Z} \mapsto \mathbb{R}^N$ is Bohr (respectively, Besicovitch) almost periodic.

6.2.1 Existence of a Bohr Almost Periodic Solution

Define the resolvent $\rho(A, B)$ by

$$\rho(A, B) := \left\{ \lambda \in \mathbb{C} : \lambda A + B \ \text{is invertible} \right\}.$$

Theorem 6.3 *If* $\mathbb{S}^1 \subseteq \rho(A, B)$, *then Eq. (6.2) has a unique almost periodic solution.*

Proof The strategy here consists of adapting our setting to that of Campbell [34]. Indeed, setting $\hat{A} = (A + B)^{-1}A$, $\hat{B} = (A + B)^{-1}B$, and $\hat{C}(t) = (A + B)^{-1}C(t)$, one can easily see that Eq. (6.2) is equivalent to,

$$\hat{A}x(t + 1) + \hat{B}x(t) = \hat{C}(t), \quad t \in \mathbb{Z}. \tag{6.3}$$

Using the identity, $\hat{A} + \hat{B} = I_N$, it follows that $\hat{A}\hat{B} = \hat{B}\hat{A}$. Consequently, one can find a common basis of trigonalization for $\hat{A}$ and $\hat{B}$. That is, there exists an invertible matrix T such that

$$\hat{A} = T^{-1} \begin{pmatrix} A_1 & 0 \\ 0 & A_2 \end{pmatrix} T, \quad \hat{B} = T^{-1} \begin{pmatrix} B_1 & 0 \\ 0 & B_2 \end{pmatrix} T,$$

where A_1, B_2 are invertible and A_2, B_1 are nilpotent.

Recall that here, $A_i + B_i = I_N$ for $i = 1, 2$. Consequently, writing

$$Tx(t) = \begin{pmatrix} w(t) \\ v(t) \end{pmatrix}$$

and

$$T\hat{C}(t) = \begin{pmatrix} \alpha(t) \\ \beta(t) \end{pmatrix},$$

where $(\alpha(t))_t$ and $(\beta(t))_t$ are almost periodic, it follows that Eq. (6.3) can be rewritten as

$$\begin{cases} A_1 w(t + 1) + B_1 w(t) = \alpha(t) \\ A_2 v(t + 1) + B_2 v(t) = \beta(t). \end{cases} \tag{6.4}$$

Using the fact that both A_1 and B_2 are invertible, one can see that Eq. (6.4) is equivalent to,

$$\begin{cases} w(t + 1) + A_1^{-1} B_1 w(t) = A_1^{-1}\alpha(t) \\ B_2^{-1} A_2 v(t + 1) + v(t) = B_2^{-1}\beta(t). \end{cases} \tag{6.5}$$

Let us now put our main focus upon the first equation appearing in Eq. (6.5), that is, the equation given by

$$w(t + 1) - (-A_1^{-1}B_1)w(t) = A_1^{-1}\alpha(t), \quad t \in \mathbb{Z}. \tag{6.6}$$

Obviously, $t \mapsto (A_1^{-1}\alpha(t))_{t \in \mathbb{Z}}$ is almost periodic. We shall now prove that $-A_1^{-1}B_1$ has no eigenvalue that belongs to $\mathbb{S}^1$. From that, we will deduce that Eq. (6.6) has a unique almost periodic solution. For that, consider a nonzero eigenvalue λ of $-A_1^{-1}B_1$. Let $x_1 \neq 0$ be an eigenvector for $-A_1^{-1}B_1$, that is, $-A_1^{-1}B_1 x_1 = \lambda x_1$. Consequently,

$$(\lambda A_1 + B_1)x_1 = 0,$$

from which we deduce that

$$(\lambda \hat{A} + \hat{B})T^{-1} \begin{pmatrix} x_1 \\ 0 \end{pmatrix} = 0.$$

Using the fact that

$$T^{-1} \begin{pmatrix} x_1 \\ 0 \end{pmatrix} \neq \begin{pmatrix} 0 \\ 0 \end{pmatrix}$$

we deduce that $\lambda \hat{A} + \hat{B}$ is not invertible, thus this is the case for $\lambda A + B$ too. With the assumption made, this proves that $|\lambda| \neq 1$. Consequently, there exists a unique almost periodic solution $(w(t))_{t \in \mathbb{Z}}$ to

$$w(t + 1) - (-A_1^{-1}B_1)w(t) = A_1^{-1}\alpha(t).$$

For the second equation appearing in Eq. (6.5), setting $V(t) = v(-t)$, it becomes, by changing t in $-t$,

$$V(t) + B_2^{-1}A_2 V(t - 1) = B_2^{-1}\beta(-t). \tag{6.7}$$

Using similar arguments as above, one can see that Eq. (6.7) has a unique almost periodic solution $(V(t))_{t \in \mathbb{Z}}$, so the second equation appearing in Eq. (6.5) has a unique almost periodic solution $(v(t))_{t \in \mathbb{Z}}$. Since Eqs. (6.5) and (6.2) are equivalent, we obtain existence and uniqueness of an almost periodic solution to Eq. (6.2). The proof is complete.

Remark 6.4 Notice that the continuous operator

$$\mathcal{T} : (x(t))_{t \in \mathbb{Z}} \to (Ax(t + 1) + Bx(t))_{t \in \mathbb{Z}}$$

is invertible and maps the Banach space of (Bohr) almost periodic sequences into itself. It follows from the *Bounded Inverse Theorem* that $\mathscr{T}^{-1}$ is also continuous. Consequently, there exists a constant $M > 0$ such that for all $(C(t))_{t \in \mathbb{Z}}$,

$$\|(x(t))_t\|_\infty \leq M \|(C(t))_t\|_\infty.$$

An immediate consequence of Theorem 6.3 is the following:

Corollary 6.5 *Let* $A = (a_{ij})$ *and* $B = (b_{ij})$ *be* $N \times N$ *square matrices and suppose that*

$$\forall i, \ \|a_{ii}| - |b_{ii}\| > \sum_{j \neq i} (|a_{ij}| + |b_{ij}|).$$

Then Eq. (6.2) has a unique almost periodic solution.

Proof Indeed, let $\lambda \in \mathbb{S}^1$ and let $c_{ij} = a_{ij}\lambda + b_{ij}$, so that $\lambda A + B = (c_{ij})$.
Now

$$|c_{ii}| = |\lambda a_{ii} + b_{ii}| \geq \|\lambda a_{ii}| - |b_{ii}\| = \|a_{ii}| - |b_{ii}\|$$

and

$$\sum_{j \neq i} |c_{ij}| = \sum_{j \neq i} |a_{ij}\lambda + b_{ij}| \leq \sum_{j \neq i} (|a_{ij}| + |b_{ij}|)$$

so thus for all i,

$$|c_{ii}| > \sum_{j \neq i} |c_{ij}|,$$

which yields $\mathbb{S}^1 \subseteq \rho(A, B)$.

In view of the above, using Theorem 6.3 it follows that Eq. (6.2) has a unique almost periodic solution. □

Remark 6.6 Let us mention that there exist infinitely many pairs of matrices (A, B) satisfying the assumption of Corollary 6.5.

6.2.2 Existence of Besicovitch Almost Periodic Solution

In this subsection, we suppose $(C(t))_{t \in \mathbb{Z}}$ is Besicovitch almost periodic and study the existence of Besicovitch almost periodic solutions to Eq. (6.2). Here, the proof is more straightforward, using tools from Fourier analysis.

Indeed, write $C(t) \sim \sum_{\alpha \in [0,2\pi)} c_\alpha \hat{e}_\alpha(t)$, where $\hat{e}_\alpha(t) = e^{i\alpha t}$ and $(c_\alpha)_\alpha \in \ell^2([0, 2\pi), \mathbb{R}^N)$. We look for a solution in the following form,

$$x(t) \sim \sum_{\alpha \in [0,2\pi)} a_\alpha \hat{e}_\alpha(t), \quad (a_\alpha)_\alpha \in \ell^2([0, 2\pi), \mathbb{R}^N).$$

Now

$$Ax(t+1) + Bx(t) \sim \sum_{\alpha \in [0,2\pi)} (\hat{e}_\alpha(1)Aa_\alpha + Ba_\alpha)\hat{e}_\alpha(t).$$

By the uniqueness of the Fourier-Bohr expansion, Eq. (6.2) is equivalent to,

$$\forall \alpha \in [0, 2\pi), \quad (\hat{e}_\alpha(1)A + B)a_\alpha = c_\alpha.$$

Since $\hat{e}_\alpha(1) \in \mathbb{S}^1$, given that $\hat{e}_\alpha(1)A + B$ is invertible, so we obtain a candidate

$$\forall \alpha \in [0, 2\pi), \quad a_\alpha = (\hat{e}_\alpha(1)A + B)^{-1}c_\alpha.$$

We need now to prove that $(a_\alpha)_\alpha \in \ell^2([0, 2\pi), \mathbb{R}^N)$. Since $\mathbb{S}^1$ is compact, then the function $\mathbb{S}^1 \mapsto (0, \infty)$, $\lambda \to \|(\lambda A + B)^{-1}\|$ is bounded and so let $M > 0$ be such that

$$\forall \lambda \in \mathbb{S}^1, \quad \|(\lambda A + B)^{-1}\| \leq M.$$

Clearly,

$$|a_\alpha|^2 \leq M^2 |c_\alpha|^2.$$

This yields $(a_\alpha)_\alpha \in \ell^2([0, 2\pi), \mathbb{R}^N)$. Further,

$$\|(x(t))_{t \in \mathbb{Z}}\|_2 \leq M \|(C(t))_{t \in \mathbb{Z}}\|_2.$$

Notice here that we have a formula for M which is given by

$$M = \sup_{\lambda \in \mathbb{S}^1} \|(\lambda A + B)^{-1}\|.$$

Remark 6.7 In the case of assumptions of Theorem 6.5, one can actually compute explicitly a bound for M. Indeed, let us consider

$$\theta := \min_{i=1,2,\ldots,N} \{\|a_{ii}\| - |b_{ii}\| - \sum_{j \neq i}(|a_{ij}| + |b_{ij}|)\}.$$

Then

$$M \leq \frac{\sqrt{n}}{\theta}.$$

Set $c_{ij} = a_{ij}\lambda + b_{ij}$. Given $\lambda \in \mathbb{S}^1$, let us consider the system $Y = (\lambda A + B)X$ and fix i_0 such that $|X_{i_0}| = \max_i |X_i|$.

Now

$$|Y|_2 \geq |Y_{i_0}| = |\sum_j (\lambda a_{i_0 j} + b_{i_0 j} X_j)| \geq$$

$$|a_{i_0 i_0}|.|X_{i_0}| - \sum_{j \neq i_0} |\lambda a_{i_0 j} + b_{i_0 j}||X_j| \geq \theta |X|_\infty,$$

thus

$$|X|_2 \leq \sqrt{n}|X|_\infty \leq \frac{\sqrt{n}}{\theta}|Y|_2.$$

We apply this with $Y = c_\alpha$ and $X = a_\alpha$.

6.3   The Semilinear Equation

First of all, note that from Sect. 6.2, we deduce that the linear operator

$$\mathcal{T} : (x(t))_{t \in \mathbb{Z}} \rightarrow (Ax(t+1) + Bx(t))_{t \in \mathbb{Z}}$$

is bijective and bi-continuous from $AP(\mathbb{Z}, \mathbb{R}^N)$ (respectively, from $B^2(\mathbb{Z}, \mathbb{R}^N)$ into itself) into itself.

Using similar arguments as above and the composition of almost periodic sequences, we can obtain the existence of Bohr (respectively, Besicovich) almost periodic solutions to Eq. (6.12).

Theorem 6.8 *Suppose $\mathbb{S}^1 \subseteq \rho(A, B)$ and that $f \in AP(\mathbb{Z}, \mathbb{R}^N)$. Further, suppose that $x \mapsto f(t, x)$ is K-Lipschitzian. Then for sufficiently small K, Eq. (6.12) has a unique Bohr almost periodic solution.*

Theorem 6.9 *Suppose $\mathbb{S}^1 \subseteq \rho(A, B)$ and that $f : b\mathbb{Z} \times \mathbb{R}^N \rightarrow \mathbb{R}^N$ is Caratheodory, $f(., 0) \in \ell^2(b\mathbb{Z}, \mathbb{R}^N)$. Further, we suppose that $x \mapsto f(t, x)$ is K-Lipschitzian. Then for sufficiently small K, Eq. (6.12) has a unique Besicovitch almost periodic solution.*

Let X be either $AP(\mathbb{Z}, \mathbb{R}^N)$ or $B^2(\mathbb{Z}, \mathbb{R}^N)$. From the assumptions upon f, the Nemytskii operator for f is given by

$$\mathcal{N}_f : ((x(t))_{t \in \mathbb{Z}}) \mapsto (f(t, x(t)))_{t \in \mathbb{Z}},$$

which maps X into itself. Moreover, $\mathcal{T}$ is bi-continuous from X to itself.

Equation (6.12) is equivalent to,

$$\mathcal{T}((x(t))_{t \in \mathbb{Z}}) = \mathcal{N}_f((x(t))_{t \in \mathbb{Z}}),$$

which is equivalent to finding a fixed point for $\mathcal{T}^{-1} \circ \mathcal{N}_f$. This nonlinear operator is $\|\mathcal{T}^{-1}\|K$-Lipschitzian. Consequently,

$$K < \|\mathcal{T}^{-1}\|^{-1}$$

to obtain the existence of a unique almost periodic solution to Eq. (6.12), we use the Banach fixed-point theorem.

6.4 Second-Order Singular Difference Equations

Of interest is the study of (respectively, Besicovitch) Bohr almost periodic to the following second-order difference equations

$$Ax(t + 2) + Bx(t + 1) + Cx(t) = f(t, x(t)), \quad t \in \mathbb{Z} \tag{6.8}$$

where A, B, C are $N \times N$-squares matrices with $\det A = \det B = \det C = 0$ and $f : \mathbb{Z} \times \mathbb{R}^N \mapsto \mathbb{R}^N$ is almost periodic in the first variable uniformly in the second one.

In order to study the existence of (respectively, Besicovitch) Bohr almost periodic solutions to Eq. (6.8), one makes extensive use of the results obtained in Sect. 6.3. For that, we rewrite Eq. (6.8) as follows:

$$Lw(t + 1) + Mw(t) = F(t, w(t)), \quad t \in \mathbb{Z} \tag{6.9}$$

where

$$L = \begin{pmatrix} B & A \\ I & O \end{pmatrix}, \quad M = \begin{pmatrix} C & O \\ O & -I \end{pmatrix}, \quad F = \begin{pmatrix} f \\ 0 \end{pmatrix}, \quad w(t) = \begin{pmatrix} x(t) \\ x(t+1) \end{pmatrix}$$

with O and I being the $N \times N$ zero and identity matrices.

Lemma 6.10 $\lambda L + M$ is invertible if and only if $\lambda(A + B) + C$ is invertible.

Proof The $2N \times 2N$ square matrix $\lambda L + M$ is given by

$$\lambda L + M = \begin{pmatrix} \lambda B + C & \lambda A \\ \lambda I & -I \end{pmatrix}.$$

Consequently, solving the system

$$(\lambda L + M) \begin{pmatrix} u \\ v \end{pmatrix} = \begin{pmatrix} x \\ y \end{pmatrix}$$

yields $(\lambda B + C)u + \lambda Av = x$ and $\lambda u - v = y$. If $\lambda^2 A + \lambda B + C$ is invertible, then from $v = \lambda u - y$ it follows that $(\lambda^2 A + \lambda B + C)u = \lambda Ay + x$ which yields,

$$u = \left[\lambda^2 A + \lambda B + C \right]^{-1} \left(\lambda Ay + x \right), \quad v = \lambda \left[\lambda^2 A + \lambda B + C \right]^{-1} \left(x + \lambda Ay \right) - y$$

which yields $\lambda L + M$ is invertible.

The proof for the converse can be done using similar arguments as above and hence is omitted.

Set

$$\rho(A, B, C) := \left\{ \lambda \in \mathbb{C} : \lambda^2 A + \lambda B + C \text{ is invertible} \right\}.$$

Using Lemma 6.10, Theorem 6.8, and Theorem 6.9, we obtain the following results:

Theorem 6.11 *Suppose* $\mathbb{S}^1 \subseteq \rho(A, B, C)$ *and that* $f \in AP(\mathbb{Z}, \mathbb{R}^N)$. *Further, suppose that* $x \mapsto f(t, x)$ *is* K-*Lipschitzian. Then for sufficiently small* K, *Eq. (6.8) has a unique Bohr almost periodic solution.*

Theorem 6.12 *Suppose* $\mathbb{S}^1 \subseteq \rho(A, B, C)$ *and that* $f : b\mathbb{Z} \times \mathbb{R}^N \to \mathbb{R}^N$ *is Caratheodory,* $f(., 0) \in \ell^2(b\mathbb{Z}, \mathbb{R}^N)$. *Further, we suppose that* $x \mapsto f(t, x)$ *is* K-*Lipschitzian. Then for sufficiently small* K, *Eq. (6.8) has a unique Besicovitch almost periodic solution.*

Let $p \geq 2$ be an integer. One should mention that the previous techniques can be easily used to study the existence of almost periodic solutions to higher order singular systems of difference equations of the form,

$$A_p x(t+p) + A_{p-1} x(t+p-1) + \ldots + A_1 x(t+1) + A_0 x(t) = f(t, x(t)), \quad (6.10)$$

for all $t \in \mathbb{Z}$, where A_k for $k = 0, 1, 2, \ldots, p$, are $N \times N$-squares matrices with $\det A_k = 0$ for $k = 0, 1, 2, .., p$, and $f : \mathbb{Z} \times \mathbb{R}^N \mapsto \mathbb{R}^N$ is almost periodic in the first variable uniformly in the second one.

Setting

$$\rho(A_p, A_{p-1}, \ldots, A_0) := \left\{\lambda \in \mathbb{C} : \lambda^p A_p + \lambda^{p-1} A_{p-1} + \ldots + \lambda A_1 + A_0 \text{ is invertible}\right\},$$

the existence results can be formulated as follows:

Theorem 6.13 *Suppose* $\mathbb{S}^1 \subseteq \rho(A_p, A_{p-1}, \ldots, A_0)$ *and that* $f \in AP(\mathbb{Z}, \mathbb{R}^N)$. *Further, suppose that* $x \mapsto f(t, x)$ *is K-Lipschitzian. Then for sufficiently small K, Eq. (6.10) has a unique Bohr almost periodic solution.*

Theorem 6.14 *Suppose* $\mathbb{S}^1 \subseteq \rho(A_p, A_{p-1}, \ldots, A_0)$ *and that* $f : b\mathbb{Z} \times \mathbb{R}^N \to \mathbb{R}^N$ *is Caratheodory,* $f(., 0) \in \ell^2(b\mathbb{Z}, \mathbb{R}^N)$. *Further, we suppose that* $x \mapsto f(t, x)$ *is K-Lipschitzian. Then for sufficiently small K, Eq. (6.10) has a unique Besicovitch almost periodic solution.*

6.5 Exercises

1. Give an example of square matrices A and B that satisfy the assumption of Corollary 6.5.
2. Prove Theorem 6.8.
3. Prove Theorem 6.9.
4. Prove Theorem 6.11.
5. Prove Theorem 6.12.
6. Prove Theorem 6.13.
7. Prove Theorem 6.14.
8. Study the existence of Bohr (respectively, Besicovitch) almost periodic solutions to the following class of nonautonomous singular difference equations,

$$A(t)x(t + 1) + B(t)x(t) = f(t, x(t)) \qquad (6.11)$$

where $f : \mathbb{Z} \times \mathbb{R}^N \to \mathbb{R}^N$ is Bohr (respectively, Besicovitch) almost periodic in $t \in \mathbb{Z}$ uniformly in the second variable, and $A(t)$, $B(t)$ are $N \times N$ square matrices satisfying $\det A(t) = \det B(t) = 0$ for all $t \in \mathbb{Z}$.

9. Study the existence of Bohr (respectively, Besicovitch) almost periodic solutions to the following class of nonautonomous singular difference equations,

$$A(t)x(t + 2) + B(t)x(t + 1) + C(t)x(t) = f(t, x(t)) \qquad (6.12)$$

where $f : \mathbb{Z} \times \mathbb{R}^N \to \mathbb{R}^N$ is Bohr (respectively, Besicovitch) almost periodic in $t \in \mathbb{Z}$ uniformly in the second variable, and $A(t)$, $B(t)$, $C(t)$ are $N \times N$ square matrices satisfying $\det A(t) = \det B(t) = \det C(t) = 0$ for all $t \in \mathbb{Z}$.

6.6 Comments

This chapter is mainly based upon the work by Diagana and Pennequin [48]. Other sources for this chapter include the work of Campbell [34]. For additional readings on the topic discussed in this chapter, we refer the reader for instance to Anh et al. [15] and Du et al. [52].

Chapter 7
Fractional Integro-Differential Equations

7.1 Introduction

Fractional calculus is a generalization of the classical differentiation and integration of non-integer order. Fractional calculus is as old as differential calculus. Fractional differential and integral equations have applications in many fields including engineering, science, finance, applied mathematics, bio-engineering, radiative transfer, neutron transport, and the kinetic theory of gases, see, e.g., [16, 33, 35, 36, 71, 74]. Noteworthy progress upon the study of ordinary and partial fractional differential equations have recently been made, see, e.g., Abbas et al. [6], Baleanu et al. [19], Diethelm [51], Kilbas et al. [77], Miller and Ross [91], Podlubny [101], and Samko et al. [103]. Further, some recent results upon the existence and attractivity of solutions to various integral equations of two variables have been obtained by many people including Abbas et al. [2, 3, 5].

In this chapter, we study the existence, uniqueness, estimates, and global asymptotic stability for some classes of fractional integro-differential equations with finite delay. In order to achieve our goal, we make extensive use of some fixed-point theorems as well as the so-called Pachpatte techniques.

Recently, Pachpatte [98] obtained some existence and uniqueness results as well as some other properties of solutions to certain Volterra integral and integro-differential equations in two variables. The main tools utilized in his analysis are based upon the applications of the Banach fixed point theorem coupled with the so-called Bielecki type norm and certain integral inequalities with explicit estimates. Using integral inequalities and a fixed-point approach, we improve some of the above-mentioned results and study the global attractivity of solutions for the system of partial fractional integro-differential equations in the form,

$$\mathbb{D}_\theta^r u(t, x) = f(t, x, u_{(t,x)}, (Gu)(t, x)), \quad \text{for } (t, x) \in J := \mathbb{R}_+ \times [0, b], \quad (7.1)$$

$$u(t, x) = \phi(t, x), \text{ if } (t, x) \in \tilde{J} := [-\alpha, \infty) \times [-\beta, b] \backslash (0, \infty) \times (0, b], \quad (7.2)$$

© Springer Nature Switzerland AG 2018 97
T. Diagana, *Semilinear Evolution Equations and Their Applications*,
https://doi.org/10.1007/978-3-030-00449-1_7

$$\begin{cases} u(t, 0) = \varphi(t); \ t \in \mathbb{R}_+, \\ u(0, x) = \psi(x); \ x \in [0, b], \end{cases} \tag{7.3}$$

where

$$(Gu)(t, x) = \frac{1}{\Gamma(r_1)\Gamma(r_2)} \int_0^t \int_0^x (t - s)^{r_1 - 1}(x - y)^{r_2 - 1} g(t, x, s, y, u_{(s,y)}) dy ds, \tag{7.4}$$

$\alpha, \beta, b > 0$, $\theta = (0, 0)$, $r = (r_1, r_2) \in (0, 1] \times (0, 1]$, $\mathbb{R}_+ = [0, \infty)$, I_θ^r is the left-sided mixed Riemann–Liouville integral of order r, D_θ^r is the standard Caputo fractional derivative of order r, $f : J \times \mathscr{C} \to \mathbb{R}$, $g : J_1 \times \mathscr{C} \to \mathbb{R}$ are given continuous functions, $J_1 := \{(t, x, s, y) : 0 \le s \le t < \infty, \ 0 \le y \le x \le b]\}$, $\varphi : \mathbb{R}_+ \to \mathbb{R}$, $\psi : [0, b] \to \mathbb{R}$ are absolutely continuous functions with $\lim_{t \to \infty} \varphi(t) = 0$, and $\psi(x) = \varphi(0)$ for each $x \in [0, b]$, $\Phi : \tilde{J} \to \mathbb{R}$ is continuous with $\varphi(t) = \Phi(t, 0)$ for each $t \in \mathbb{R}_+$, and $\psi(x) = \Phi(0, x)$ for each $x \in [0, b]$, $\Gamma(.)$ is the Gamma function defined by

$$\Gamma(\xi) = \int_0^\infty t^{\xi - 1} e^{-t} dt; \ \xi > 0,$$

and $\mathscr{C} := C([-\alpha, 0] \times [-\beta, 0])$ is the space of continuous functions on $[-\alpha, 0] \times [-\beta, 0]$ with the standard norm

$$\|u\|_{\mathscr{C}} = \sup_{(t,x) \in [-\alpha, 0] \times [-\beta, 0]} |u(t, x)|.$$

If $u \in C := C([-\alpha, \infty) \times [-\beta, b])$, then for any $(t, x) \in J$ define $u_{(t,x)}$ by

$$u_{(t,x)}(\tau, \xi) = u(t + \tau, x + \xi); \ \text{for} \ (\tau, \xi) \in [-\alpha, 0] \times [-\beta, 0].$$

7.2 Preliminaries and Notations

Let $a, b > 0$ and $L^1([0, a] \times [0, b])$ be the space of Lebesgue-integrable functions $u : [0, a] \times [0, b] \to \mathbb{R}$ equipped with the norm,

$$\|u\|_1 = \int_0^a \int_0^b |u(t, x)| dx dt.$$

By $C := C(J)$ we denote the space of all continuous functions from J into $\mathbb{R}$. Similarly, by $BC := BC([-\alpha, \infty) \times [-\beta, b])$ we denote the Banach space of all bounded and continuous functions from $[-\alpha, \infty) \times [-\beta, b]$ into $\mathbb{R}$ equipped with the standard sup norm which we denoted by

$$\|u\|_{BC} = \sup_{(t,x) \in [-\alpha, \infty) \times [-\beta, b]} |u(t, x)|.$$

Definition 7.1 For $u_0 \in BC$ and $\eta \in (0, \infty)$, we denote by $B(u_0, \eta)$, the closed ball in BC centered at u_0 with radius η.

Definition 7.2 ([108]) Let $r = (r_1, r_2) \in (0, \infty) \times (0, \infty)$, $\theta = (0, 0)$ and $u \in L^1([0, a] \times [0, b])$. The left-sided mixed Riemann–Liouville integral of order r of u is defined by

$$(I_\theta^r u)(t, x) = \frac{1}{\Gamma(r_1)\Gamma(r_2)} \int_0^t \int_0^x (t - s)^{r_1 - 1}(x - y)^{r_2 - 1}u(s, y)dyds.$$

In particular,

$$(I_\theta^\theta u)(t, x) = u(t, x), \quad (I_\theta^\sigma u)(t, x) = \int_0^t \int_0^x u(s, y)dyds;$$

$$\text{for almost all } (t, x) \in [0, a] \times [0, b],$$

where $\sigma = (1, 1)$.
For instance, $I_\theta^r u$ exists for all $r_1, r_2 > 0$, when $u \in L^1([0, a] \times [0, b])$. Moreover

$$(I_\theta^r u)(t, 0) = (I_\theta^r u)(0, x) = 0; \quad t \in [0, a], \ x \in [0, b].$$

Example 7.3 Let $\lambda, \omega \in (-1, 0) \cup (0, \infty)$ and $r = (r_1, r_2) \in (0, \infty) \times (0, \infty)$, then

$$I_\theta^r t^\lambda x^\omega = \frac{\Gamma(1 + \lambda)\Gamma(1 + \omega)}{\Gamma(1 + \lambda + r_1)\Gamma(1 + \omega + r_2)}t^{\lambda + r_1}x^{\omega + r_2},$$

$$\text{for almost all } (t, x) \in [0, a] \times [0, b].$$

By $1 - r$ we mean $(1 - r_1, 1 - r_2) \in [0, 1) \times [0, 1)$. Denote by $D_{tx}^2 := \frac{\partial^2}{\partial t \partial x}$, the mixed second order partial derivative.

Definition 7.4 ([108]) Let $r \in (0, 1] \times (0, 1]$ and $u \in L^1([0, a] \times [0, b])$. Recall that the Caputo fractional derivative of order r of u is defined by the expression

$$\mathbb{D}_\theta^r u(t, x) = (I_\theta^{1-r} D_{tx}^2 u)(t, x)$$

$$= \frac{1}{\Gamma(1 - r_1)\Gamma(1 - r_2)} \int_0^t \int_0^x \frac{(D_{sy}^2 u)(s, y)}{(t - s)^{r_1}(x - y)^{r_2}}dyds.$$

The case when $\sigma = (1, 1)$ is included and we have

$$(\mathbb{D}_\theta^\sigma u)(t, x) = (D_{xy}^2 u)(t, x), \text{ for almost all } (t, x) \in [0, a] \times [0, b].$$

Example 7.5 Let $\lambda, \omega \in (-1, 0) \cup (0, \infty)$ and $r = (r_1, r_2) \in (0, 1] \times (0, 1]$, then

$$\mathbb{D}_\theta^r t^\lambda x^\omega = \frac{\Gamma(1+\lambda)\Gamma(1+\omega)}{\Gamma(1+\lambda-r_1)\Gamma(1+\omega-r_2)} t^{\lambda-r_1} x^{\omega-r_2},$$

for almost all $(t, x) \in [0, a] \times [0, b]$.

In the sequel, we need the following lemma

Lemma 7.6 ([1]) *Let $f \in L^1([0, a] \times [0, b])$. A function $u \in AC([0, a] \times [0, b])$ is a solution to the problem*

$$\begin{cases} (\mathbb{D}_\theta^r u)(t, x) = f(t, x); \ (t, x) \in [0, a] \times [0, b], \\ u(t, 0) = \varphi(t); \ t \in [0, a], \ u(0, x) = \psi(x); \ x \in [0, b], \\ \varphi(0) = \psi(0), \end{cases}$$

if and only if u satisfies

$$u(t, x) = \mu(t, x) + (I_\theta^r f)(t, x); \ (t, x) \in [0, a] \times [0, b],$$

where

$$\mu(t, x) = \varphi(t) + \psi(x) - \varphi(0).$$

Denote by $D_1 := \frac{\partial}{\partial t}$, the partial derivative of a function defined on J (or J_1) with respect to the first variable, $D_2 := \frac{\partial}{\partial x}$, $D_2 D_1 := \frac{\partial^2}{\partial t \partial x}$. In the sequel we will make use of the following Lemma due to Pachpatte.

Lemma 7.7 ([98]) *Let $u, e, p \in C(J)$, $k, D_1 k, D_2 k, D_2 D_1 k \in C(J_1)$ be positive functions. If $e(t, x)$ is nondecreasing in each variable $(t, x) \in J$ and*

$$u(t, x) \le e(t, x) + \int_0^t \int_0^x p(s, y)$$

$$\times \left[u(s, y) + \int_0^s \int_0^y k(s, y, \tau, \xi) u(\tau, \xi) d\xi d\tau \right] dy ds; \ (t, x) \in J, \quad (7.5)$$

then,

$$u(t, x) \le e(t, x) \left[1 + \int_0^t \int_0^x p(s, y) \exp\left(\int_0^s \int_0^y [p(\tau, \xi) + A(\tau, \xi)] d\xi d\tau \right) dy ds \right]$$

$$(7.6)$$

for all $(t, x) \in J$, where

$$A(t, x) = k(t, x, s, y) + \int_0^t D_1 k(t, x, s, y)ds + \int_0^x D_2 k(t, x, s, y)dy$$

$$+ \int_0^t \int_0^x D_2 D_1 k(t, x, s, y)dyds; \quad (t, x) \in J. \tag{7.7}$$

Let G be an operator from $\emptyset \neq \Omega \subset BC$ into itself and consider the solutions of equation

$$(Gu)(t, x) = u(t, x). \tag{7.8}$$

Now we review the concept of attractivity of solutions to Eq. (7.8). For $u_0 \in BC$ and $\eta \in (0, \infty)$, we denote by $B(u_0, \eta)$, the closed ball in BC centered at u_0 with radius η.

Definition 7.8 ([5]) Solutions to Eq. (7.8) are locally attractive if there exists a ball $B(u_0, \eta)$ in the space BC such that for any arbitrary solutions $v = v(t, x)$ and $w = w(t, x)$ to Eq. (7.8) belonging to $B(u_0, \eta) \cap \Omega$, we have that, for each $x \in [0, b]$,

$$\lim_{t \to \infty} (v(t, x) - w(t, x)) = 0. \tag{7.9}$$

Definition 7.9 When the limit to Eq. (7.9) is uniform with respect to $B(u_0, \eta)$, solutions to Eq. (7.8) are said to be locally attractive (or equivalently that solutions to Eq. (7.8) are asymptotically stable).

Definition 7.10 ([5]) The solution $v = v(t, x)$ of Eq. (7.8) is said to be globally attractive if Eq. (7.9) holds for each solution $w = w(t, x)$ of Eq. (7.8). If condition Eq. (7.9) is satisfied uniformly with respect to the set Ω, solutions of Eq. (7.8) are said to be globally asymptotically stable (or uniformly globally attractive).

7.3 Main Results

Prior to getting into technical considerations and estimates, let us define what we mean by a solution to the system Eqs. (7.1)–(7.3).

Definition 7.11 A function $u \in BC$ whose mixed derivative D_{tx}^2 exists and is integrable, is said to be a solution to the system Eqs. (7.1)–(7.3), if u satisfies Eqs. (7.1) and (7.3) on J and that Eq. (7.2) on $\tilde{J}$ holds.

7.3.1 Existence and Uniqueness

Our first result is devoted to the existence and uniqueness of a solution to Eqs. (7.1)–(7.3).

Theorem 7.12 *Assume that the following assumptions hold,*

(H.61)   The function φ is continuous and bounded with

$$\varphi^* = \sup_{(t,x)\in\mathbb{R}_+\times[0,b]} |\varphi(t,x)|;$$

(H.62)   There exist positive functions $p_1, p_2 \in BC(J)$ such that

$$|f(t, x, u_1, u_2) - f(t, x, v_1, v_2)| \leq p_1(t, x)\|u_1 - v_1\|_{\mathscr{C}} + p_2(t, x)|u_2 - v_2|,$$

for each $(t, x) \in J$, $u_1, v_1 \in \mathscr{C}$ and $u_2, v_2 \in \mathbb{R}$. Moreover, assume that the function

$$t \to \int_0^t \int_0^x (t-s)^{r_1-1}(x-y)^{r_2-1} f(s, y, 0, (G0)(s, y)) \, dy \, ds$$

is bounded on J with

$$f^* = \sup_{(t,x)\in J} \frac{1}{\Gamma(r_1)\Gamma(r_2)} \int_0^t \int_0^x (t-s)^{r_1-1}(x-y)^{r_2-1}|f(s, y, 0, (G0)(s, y))| \, dy \, ds;$$

(H.63)   There exists a positive function $q \in BC(J_1)$ such that

$$|g(t, x, s, y, u) - g(t, x, s, y, v)| \leq q(t, x, s, y)|u - v|,$$

for each $(t, x, s, y) \in J_1$ and $u, v \in \mathbb{R}$.

If

$$p_1^* + p_2^* q^* < 1, \tag{7.10}$$

where

$$p_i^* = \sup_{(t,x)\in J} \left[\frac{1}{\Gamma(r_1)\Gamma(r_2)} \int_0^t \int_0^x (t-s)^{r_1-1}(x-y)^{r_2-1} p_i(s, y) \, dy \, ds \right]; \quad i = 1, 2,$$

and

$$q^* = \sup_{(t,x)\in J} \left[\frac{1}{\Gamma(r_1)\Gamma(r_2)} \int_0^t \int_0^x (t-s)^{r_1-1}(x-y)^{r_2-1} q(t, x, s, y) \, dy \, ds \right],$$

then the system (7.1)–(7.3) has a unique solution on $[-\alpha, \infty) \times [-\beta, b]$.

Proof Define the nonlinear operator $N : BC \to BC$ by

$$(Nu)(t, x) = \begin{cases} \Phi(t, x), & (t, x) \in \tilde{J}, \\ \\ \varphi(t) + \dfrac{1}{\Gamma(r_1)\Gamma(r_2)} \displaystyle\int_0^t \int_0^x (t - s)^{r_1 - 1}(x - y)^{r_2 - 1} \\ \quad \times f(s, y, u_{(s,y)}, (Gu)(s, y))dyds, & (t, x) \in J. \end{cases}$$

$$(7.11)$$

Clearly, the function $(t, x) \mapsto (Nu)(t, x)$ is continuous on $[-\alpha, \infty) \times [-\beta, b]$. The next step consists of showing that $N(u) \in BC$ for each $u \in BC$. Indeed, for each $(t, x) \in \tilde{J}$, we have

$$|\Phi(t, x)| \leq \sup_{(t,x)\in\tilde{J}} |\Phi(t, x)| := \Phi^*,$$

and so $\Phi \in BC$.

From (H.62), and for arbitrarily fixed $(t, x) \in J$, we have

$$|(Nu)(t, x)| = \left| \varphi(t) + \frac{1}{\Gamma(r_1)\Gamma(r_2)} \int_0^t \int_0^x (t - s)^{r_1 - 1}(x - y)^{r_2 - 1} \right.$$

$$\left. \times f(s, y, u_{(s,y)}, (Gu)(s, y))dyds \right|$$

$$\leq |\varphi(t)| + \frac{1}{\Gamma(r_1)\Gamma(r_2)} \int_0^t \int_0^x (t - s)^{r_1 - 1}(x - y)^{r_2 - 1}$$

$$|f(s, y, u_{(s,y)}, (Gu)(s, y)) - f(s, y, 0, (G0)(s, y))|dyds$$

$$+ \frac{1}{\Gamma(r_1)\Gamma(r_2)} \int_0^t \int_0^x (t - s)^{r_1 - 1}(x - y)^{r_2 - 1}|f(s, y, 0, (G0)(s, y))|dyds$$

$$\leq |\varphi(t)| + \frac{1}{\Gamma(r_1)\Gamma(r_2)} \int_0^t \int_0^x (t - s)^{r_1 - 1}(x - y)^{r_2 - 1}$$

$$\times \left(p_1(s, y)|u_{(s,y)}| + p_2(s, y)|(Gu)(s, y)| \right) dyds$$

$$+ \frac{1}{\Gamma(r_1)\Gamma(r_2)} \int_0^t \int_0^x (t - s)^{r_1 - 1}(x - y)^{r_2 - 1}|f(s, y, 0, (G0)(s, y))|dyds$$

$$\leq \varphi^* + f^* + p_1^*\|u\|_{BC} + \frac{1}{\Gamma(r_1)\Gamma(r_2)} \int_0^t \int_0^x (t - s)^{r_1 - 1}(x - y)^{r_2 - 1}$$

$$\times p_2(s, y)|(Gu)(s, y) - (G0)(s, y)|dyds.$$

$$(7.12)$$

Now (H.63) yields

$$|(Gu)(t, x) - (G0)(t, x)|$$

$$\leq \frac{1}{\Gamma(r_1)\Gamma(r_2)} \int_0^t \int_0^x (t-s)^{r_1-1}(x-y)^{r_2-1}$$

$$\times |g(t, x, s, y, u(s, y)) - g(t, x, s, y, 0)|dyds$$

$$\leq \frac{1}{\Gamma(r_1)\Gamma(r_2)} \int_0^t \int_0^x (t-s)^{r_1-1}(x-y)^{r_2-1}q(t, x, s, y)|u(s, y)|dyds$$

$$\leq q^* \|u\|_{BC}.$$

From (7.12) we get

$$|(Nu)(t, x)| \leq \varphi^* + f^* + p_1^* \|u\|_{BC}$$

$$+ \frac{q^* \|u\|_{BC}}{\Gamma(r_1)\Gamma(r_2)} \int_0^t \int_0^x (t-s)^{r_1-1}(x-y)^{r_2-1}p_2(s, y)dyds$$

$$\leq \varphi^* + f^* + p_1^* \|u\|_{BC} + p_2^* q^* \|u\|_{BC}$$

$$\leq \varphi^* + f^* + (p_1^* + p_2^* q^*)\|u\|_{BC}.$$

thus $N(u) \in BC$.

Let $u, v \in BC$. Using our assumptions, for each $(t, x) \in J$, we obtain,

$$|(Nu)(t, x) - (Nv)(t, x)| \leq \frac{1}{\Gamma(r_1)\Gamma(r_2)} \int_0^t \int_0^x (t-s)^{r_1-1}(x-y)^{r_2-1}$$

$$\times |f(s, y, u_{(s,y)}, (Gu)(s, y)) - f(s, y, v_{(s,y)}, (Gv)(s, y))|dyds$$

$$\leq \frac{1}{\Gamma(r_1)\Gamma(r_2)} \int_0^t \int_0^x (t-s)^{r_1-1}(x-y)^{r_2-1}$$

$$\times (p_1(s, y)\|u_{(s,y)} - v_{(s,y)}\|_{\mathscr{C}} + p_2(s, y)|(Gu)(s, y) - (Gv)(s, y)|)dyds$$

$$\leq \frac{\|u - v\|_{BC}}{\Gamma(r_1)\Gamma(r_2)} \int_0^t \int_0^x (t-s)^{r_1-1}(x-y)^{r_2-1}p_1(s, y)dyds$$

$$+ \frac{\|u - v\|_{BC}}{\Gamma^2(r_1)\Gamma^2(r_2)} \int_0^t \int_0^x (t-s)^{r_1-1}(x-y)^{r_2-1}$$

$$\times p_2(s, y)\left(\int_0^s \int_0^y (s-\tau)^{r_1-1}(y-\xi)^{r_2-1}q(s, t, \tau, \xi)d\xi d\tau \right) dyds$$

$$\leq (p_1^* + p_2^* q^*)\|u - v\|_{BC}.$$

From Eq. (7.10), it follows from the Banach fixed-point principle (Theorem 1.30) that N has a unique fixed point in BC which is the solution to Eqs. (7.1)–(7.3).

7.3.2   Estimates for Solutions

Theorem 7.13 *Set*

$$d = \varphi^* + f^*. \tag{7.13}$$

Suppose that assumptions (H.61)–(H.63) hold and that we have,

(H.64) $p_1 = p_2$ and there exists a positive function $p \in BC(J)$ such that,

$$p_1(s, y) \le \Gamma(r_1)\Gamma(r_2)(t - s)^{1-r_1}(x - y)^{1-r_2} p(s, y), \text{ for each } (t, x, s, y) \in J_1,$$

(H.65) $k, D_1 k, D_2 k, D_2 D_1 k \in BC(J_1)$, where

$$k(t, x, s, y) = \frac{1}{\Gamma(r_1)\Gamma(r_2)}(t - s)^{r_1-1}(x - y)^{r_2-1} q(t, x, s, y).$$

For any solution u to Eqs. (7.1)–(7.3) on $[-\alpha, \infty) \times [-\beta, b]$, then for each $(t, x) \in J$,

$$|u(t, x)| \le d \left[1 + \int_0^t \int_0^x p(s, y) \exp\left(\int_0^s \int_0^y [p(\tau, \xi) + A(\tau, \xi)] d\xi d\tau \right) dy ds \right],$$
$$\tag{7.14}$$

where $A(t, x)$ is defined by Eq. (7.7).

Proof Using the fact that u is a solution to Eqs. (7.1)–(7.3) and from our assumptions, we have, for each $(t, x) \in J$,

$|u(t, x)| \le |\varphi(t)|$

$$+ \frac{1}{\Gamma(r_1)\Gamma(r_2)} \int_0^t \int_0^x (t - s)^{r_1-1}(x - y)^{r_2-1} |f(t, x, 0, (G0)(t, x))| dy ds$$

$$+ \frac{1}{\Gamma(r_1)\Gamma(r_2)} \int_0^t \int_0^x (t - s)^{r_1-1}(x - y)^{r_2-1}$$

$$\times |f(s, y, u_{(s,y)}, (Gu)(s, y)) - f(s, y, 0, (G0)(s, y))| dy ds$$

$$\le \varphi^* + f^* + \frac{1}{\Gamma(r_1)\Gamma(r_2)} \int_0^t \int_0^x (t - s)^{r_1-1}(x - y)^{r_2-1} p_1(s, y) \left[\|u_{(s,y)}\|_{\mathscr{C}} \right.$$

$$\left. + \frac{1}{\Gamma(r_1)\Gamma(r_2)} \int_0^s \int_0^y (s - \tau)^{r_1-1}(y - \xi)^{r_2-1} q(s, y, \tau, \xi) |u(\tau, \xi)| d\xi d\tau \right] dy ds$$

$$\leq d + \int_0^t \int_0^x p(s, y) \left[\|u_{(s,y)}\|_{\mathscr{C}} \right.$$

$$+ \frac{1}{\Gamma(r_1)\Gamma(r_2)} \int_0^s \int_0^y q(s, y, \tau, \xi) |u(\tau, \xi)| d\xi d\tau \right] dy ds$$

$$\leq d + \int_0^t \int_0^x p(s, y) \left[\|u_{(s,y)}\|_{\mathscr{C}} + \int_0^s \int_0^y k(s, y, \tau, \xi) |u(\tau, \xi)| d\xi d\tau \right] dy ds.$$

Consider the function w defined by

$$w(t, x) = \sup\{\|u(s, y)\| : -\alpha \leq s \leq t, \ -\beta \leq y \leq x\}, \ 0 \leq t < \infty, \ 0 \leq x \leq b.$$

Let $(t^*, x^*) \in [-\alpha, t] \times [-\beta, x]$ be such that $w(t, x) = |u(t^*, x^*)|$. If $(t^*, x^*) \in \tilde{J}$, then $w(t, x) = \|\Phi\|_{\mathscr{C}}$ and the previous inequality holds. If $(t^*, x^*) \in J$, then by the previous inequality, we have for $(t, x) \in J$,

$$w(t, x) \leq d + \int_0^t \int_0^x p(s, y) \left[w(s, y) + \int_0^s \int_0^y k(s, y, \tau, \xi) w(\tau, \xi) d\xi d\tau \right] dy ds.$$

From Lemma 7.7, we get

$$w(t, x) \leq d \left[1 + \int_0^t \int_0^x p(s, y) \exp\left(\int_0^s \int_0^y [p(\tau, \xi) + A(\tau, \xi)] d\xi d\tau \right) dy ds \right];$$

$$(t, x) \in J. \tag{7.15}$$

But, for every $(t, x) \in J$, $\|u_{(t,x)}\|_{\mathscr{C}} \leq w(t, x)$. Hence, Eq. (7.15) yields Eq. (7.14).

Theorem 7.14 *Set*

$$\overline{d} := f^* + \varphi^* p^* (1 + q^*). \tag{7.16}$$

Suppose that assumptions (H.61)–(H.65) hold. For any solution u to Eq. (7.2) on $[-\alpha, \infty) \times [-\beta, b]$, we have the following estimates,

$$|u(t, x) - \varphi(t)|$$

$$\leq \overline{d} \left[1 + \int_0^t \int_0^x p(s, y) \exp\left(\int_0^s \int_0^y [p(\tau, \xi) + A(\tau, \xi)] d\xi d\tau \right) dy ds \right] \tag{7.17}$$

for all $(t, x) \in J$, where A is given by Eq. (7.7).

Proof Let $h(t, x) = |u(t, x) - \varphi(t)|$. Using the fact that u is a solution to Eqs. (7.1)–(7.3) combined with our assumptions, it follows that, for each $(t, x) \in J$,

$$h(t, x) \leq \frac{1}{\Gamma(r_1)\Gamma(r_2)} \int_0^t \int_0^x (t-s)^{r_1-1}(x-y)^{r_2-1}$$

$$\times |f(s, y, u_{(s,y)}, (Gu)(s, y)) - f(s, y, \varphi(s), (G\varphi)(s))| dy ds$$

$$+ \frac{1}{\Gamma(r_1)\Gamma(r_2)} \int_0^t \int_0^x (t-s)^{r_1-1}(x-y)^{r_2-1}|f(s, y, \varphi(s), (G\varphi)(s))| dy ds$$

$$\leq \overline{d} + \frac{1}{\Gamma(r_1)\Gamma(r_2)} \int_0^t \int_0^x (t-s)^{r_1-1}(x-y)^{r_2-1}$$

$$\times |f(s, y, u_{(s,y)}, (Gu)(s, y)) - f(s, y, \varphi(s), (G\varphi)(s))| dy ds$$

$$\leq \overline{d} + \int_0^t \int_0^x (t-s)^{r_1-1}(x-y)^{r_2-1} p(s, y)$$

$$\times \left[h(s, y) + \int_0^s \int_0^y k(s, y, \tau, \xi) h(\tau, \xi) d\xi d\tau \right] dy ds. \tag{7.18}$$

Using Lemma 7.7 and Eq. (7.18), one obtains Eq. (7.17).

7.3.3 Global Asymptotic Stability of Solutions

Our main objective here is to study the global asymptotic stability of solution. For that, we show that under more suitable conditions on the functions involved in Eqs. (7.1)–(7.3) that the solutions go zero exponentially as $t \to \infty$.

Theorem 7.15 *Suppose that assumptions (H.64)–(H.65) hold and that*

(H.66)   There exist constants $\lambda > 0$ and $M \geq 0$ such that

$$|\varphi(t)| \leq M e^{-\lambda t}; \tag{7.19}$$

$$|f(t, x, u_1, u_2) - f(t, x, v_1, v_2)| \leq p_1(t, x) e^{-\lambda t} (\|u_1 - v_1\|_\mathscr{C} + |u_2 - v_2|), \tag{7.20}$$

for each $(t, x) \in J$, $u_1, v_1 \in \mathscr{C}$, $u_2, v_2 \in \mathbb{R}$,

$$|g(t, x, s, y, u) - g(t, x, s, y, v)| \leq q(t, x, s, y)|u - v|; \tag{7.21}$$

for each $(t, x, s, y) \in J_1$, $u, v \in \mathbb{R}$, and $f(t, x, 0, (G0)(t, x)) = 0$; for each $(t, x) \in J$ and the functions p, q be as in Theorem 7.13; and

(H.67) $\int_0^\infty \int_0^x [p(s, y) + A(s, y)] dy ds < \infty$, where A is given by Eq. (7.7).

If u is any solution of Eqs. (7.1)–(7.3) on $[-\alpha, \infty) \times [-\beta, b]$, then all solutions to Eqs. (7.1)–(7.3) are uniformly globally attractive on J.

Proof From our assumptions, we have, for each $(t, x) \in J$,

$$
|u(t, x)| \le |\varphi(t)| + \frac{1}{\Gamma(r_1)\Gamma(r_2)} \int_0^t \int_0^x (t - s)^{r_1-1}(x - y)^{r_2-1}
$$
$$
\times |f(s, y, u_{(s,y)}, (Gu)(s, y)) - g(s, y, 0, (G0)(s, y))| dy ds
$$
$$
+ \frac{1}{\Gamma(r_1)\Gamma(r_2)} \int_0^t \int_0^x (t - s)^{r_1-1}(x - y)^{r_2-1}|f(s, y, 0, (G0)(s, y))| dy ds
$$
$$
\le M e^{-\lambda t} + \int_0^t \int_0^x p(s, y)e^{-\lambda t}\left[u_{(s,y)} + \frac{1}{\Gamma(r_1)\Gamma(r_2)}\right.
$$
$$
\left. \times \int_0^s \int_0^y (s - \tau)^{r_1-1}(y - \xi)^{r_2-1}q(s, y, \tau, \xi)|u(\tau, \xi)|d\xi d\tau\right] dy ds. \qquad (7.22)
$$

From Eq. (7.22), we get

$$
|u(t, x)|e^{\lambda t} \le M + \int_0^t \int_0^x p(s, y)\left[u_{(s,y)} + k(s, y, \tau, \xi)|u(\tau, \xi)|d\xi d\tau\right] dy ds.
$$
$$
(7.23)
$$

Using Lemma 7.7 to Eq. (7.23) we obtain

$$
|u(t, x)|e^{\lambda t} \le M\left[1 + \int_0^t \int_0^x p(s, y)\exp\left(\int_0^s \int_0^y [p(\tau, \xi) + A(\tau, \xi)]d\xi d\tau\right) dy ds\right];
$$
$$
(t, x) \in J, \qquad (7.24)
$$

Multiplying both sides of Eq. (7.24) by $e^{-\lambda t}$ and in view of (H. 66), we get

$$
|u(t, x)| \le M\left[e^{-\lambda t} + \int_0^t \int_0^x p(s, y)\right.
$$
$$
\left. \exp\left(-\lambda t + \int_0^s \int_0^y [p(\tau, \xi) + A(\tau, \xi)]d\xi d\tau\right) dy ds\right].
$$

Thus, for each $x \in [0, b]$,

$$
\lim_{t \to \infty} u(t, x) = 0.
$$

Therefore, the solution u goes to zero as $t \to \infty$. Consequently, all solutions to Eqs. (7.1)–(7.3) are uniformly globally attractive on $[-\alpha, \infty) \times [-\beta, b]$.

7.4 Example

To illustrate our previous results, we consider the system of partial fractional integro-differential equations of the form,

$$\mathbb{D}_\theta^r u(t, x) = f(t, x, u_{(t,x)}, (Gu)(t, x)); \quad \text{for } (t, x) \in J := \mathbb{R}_+ \times [0, 1], \tag{7.25}$$

$$u(t, x) = \frac{1}{1 + t^2}; \text{ if } (t, x) \in \tilde{J} := [-1, \infty) \times [-2, 1] \backslash (0, \infty) \times (0, 1], \tag{7.26}$$

$$\begin{cases} u(t, 0) = \dfrac{1}{1 + t^2}; \ t \in \mathbb{R}_+, \\ u(0, x) = 1; \ x \in [0, 1], \end{cases} \tag{7.27}$$

where

$$(Gu)(t, x) = \frac{1}{\Gamma(r_1)\Gamma(r_2)} \int_0^t \int_0^x (t - s)^{r_1 - 1}(x - y)^{r_2 - 1} g(t, x, s, y, u_{(s,y)}) \, dy \, ds, \tag{7.28}$$

$r_1, r_2 \in (0, 1]$,

$$\begin{cases} f(t, x, u, v) = \dfrac{x^2 t^{-r_1} \sin t}{2c(1 + t^{-\frac{1}{2}})(1 + |u(t + 1, x + 2)| + |v|)}; \\ \quad for \ (t, x) \in J, \ t \neq 0 \ and \ u \in \mathscr{C}, \ v \in \mathbb{R}, \\ f(0, x, u, v) = 0, \end{cases}$$

$$c := \frac{\Gamma(\frac{1}{2})}{\Gamma(\frac{1}{2} + r_1)} \left(1 + \frac{\Gamma(\frac{1}{2})e}{\Gamma(\frac{1}{2} + r_1)\Gamma(1 + r_2)} \right),$$

$$\begin{cases} g(t, x, s, y, u) = \dfrac{t^{-r_1} s^{-\frac{1}{2}} e^{x - y - \frac{1}{s} - \frac{1}{t}}}{2c(1 + t^{-\frac{1}{2}})(1 + |u|)}; \ for \ (t, x, s, y) \in J_1, \ st \neq 0 \ and \ u \in \mathbb{R}, \\ g(t, x, 0, y, u) = g(0, x, s, y, u) = 0, \end{cases}$$

and

$$J_1 = \{(t, x, s, y) : 0 \leq s \leq t < \infty, \ 0 \leq y \leq x \leq 1\}.$$

Set

$$\varphi(t) = \frac{1}{1 + t^2}; \ t \in \mathbb{R}_+.$$

One can see that (H.61) holds as the function φ is continuous and bounded with $\varphi^* = 1$.

For each $u_1, v_1 \in \mathscr{C}$, $u_2, v_2 \in \mathbb{R}$ and $(t, x) \in J$, we have

$$|f(t, x, u_1, u_2) - f(t, x, s, v_1, v_2)|$$

$$\leq \frac{1}{2c(1 + t^{-\frac{1}{2}})} \left(x^2 t^{-r_1} |\sin t| \right) (|u_1 - v_1| + |u_2 - v_2|),$$

and for each $u, v \in \mathbb{R}$ and $(t, x, s, y) \in J_1$, we have

$$|g(t, x, s, y, u) - g(t, x, s, y, v)| \leq \frac{1}{2c(1 + t^{-\frac{1}{2}})} \left(t^{-r_1} s^{-\frac{1}{2}} e^{x - y - t - \frac{1}{s} - \frac{1}{t}} \right) |u - v|.$$

Therefore, (H.62) holds with

$$\begin{cases} p_1(t, x) = p_2(t, x) = \dfrac{x^2 t^{-r_1} |\sin t|}{2c(1 + t^{-\frac{1}{2}})}; \ t \neq 0, \\[2mm] p_1(0, x) = p_2(0, x) = 0, \end{cases}$$

and assumption (H.63) holds with

$$\begin{cases} q(t, x, s, y) = \dfrac{1}{2c(1 + t^{-\frac{1}{2}})} \left(t^{-r_1} s^{-\frac{1}{2}} e^{x - y - t - \frac{1}{s} - \frac{1}{t}} \right); \ st \neq 0, \\[2mm] q(t, x, 0, y) = k(0, x, 0, y) = 0. \end{cases}$$

We shall show that Eq. (7.10) holds with $b = 1$. Indeed,

$$\frac{1}{\Gamma(r_1)\Gamma(r_2)} \int_0^t \int_0^x (t - s)^{r_1 - 1} (x - y)^{r_2 - 1} p_1(s, y) dy ds$$

$$\leq \frac{1}{2c(1 + t^{-\frac{1}{2}})\Gamma(r_1)\Gamma(r_2)} \int_0^t \int_0^1 (t - s)^{r_1 - 1}(1 - y)^{r_2 - 1} x^2 t^{-r_1} dy ds$$

$$\leq \frac{\Gamma(\frac{1}{2}) e t^{-\frac{1}{2}}}{2c(1 + t^{-\frac{1}{2}})\Gamma(\frac{1}{2} + r_1)\Gamma(1 + r_2)},$$

then

$$p_1^* = p_2^* \leq \frac{\Gamma(\frac{1}{2})}{2c\Gamma(\frac{1}{2} + r_1)}.$$

Now

$$\frac{1}{\Gamma(r_1)\Gamma(r_2)} \int_0^t \int_0^x (t-s)^{r_1-1}(x-y)^{r_2-1} q(t,x,s,y) dy ds$$

$$\leq \frac{1}{2c(1+t^{-\frac{1}{2}})\Gamma(r_1)\Gamma(r_2)} \int_0^t \int_0^1 (t-s)^{r_1-1}(1-y)^{r_2-1} t^{-r_1} s^{-\frac{1}{2}} e^x dy ds$$

$$\leq e^x t^{-r_1} t^{-\frac{1}{2}+r_1} \frac{\Gamma(\frac{1}{2})}{2c(1+t^{-\frac{1}{2}})\Gamma(\frac{1}{2}+r_1)\Gamma(1+r_2)}$$

$$\leq \frac{\Gamma(\frac{1}{2}) e t^{-\frac{1}{2}}}{2c(1+t^{-\frac{1}{2}})\Gamma(\frac{1}{2}+r_1)\Gamma(1+r_2)},$$

then

$$q^* \leq \frac{e\Gamma(\frac{1}{2})}{2c\Gamma(\frac{1}{2}+r_1)\Gamma(1+r_2)}.$$

Thus,

$$p_1^* + p_2^* q^* \leq \frac{\Gamma(\frac{1}{2})}{2c\Gamma(\frac{1}{2}+r_1)} \left(1 + \frac{\Gamma(\frac{1}{2})e}{\Gamma(\frac{1}{2}+r_1)\Gamma(1+r_2)}\right) = \frac{1}{2} < 1,$$

which holds for each $r_1, r_2 \in (0, \infty)$. Consequently Theorem 7.12 yields Eq. (7.25)—(7.27) has a unique solution defined on $[-1, \infty) \times [-2, 1]$.

7.5 Comments

This chapter is mainly based upon the following source: Abbas et al. [7] with some slight modifications. For additional readings on similar topics and related issues, we refer the readers to the following references: [16, 33, 35, 36, 71, 74]. Furthermore, recent progress made upon the study of ordinary and partial fractional differential equations can be found in the following books: Abbas et al. [6], Baleanu et al. [19], Diethelm [51], Kilbas et al. [77], Miller and Ross [91], Podlubny [101], and Samko et al. [103].

Chapter 8
First-Order Semilinear Evolution Equations

In this chapter we study and establish the existence of classical and (bounded and almost periodic) mild solutions to some semilinear evolutions including nonautonomous ones.

8.1 First-Order Autonomous Evolution Equations

8.1.1 Existence of Mild and Classical Solutions

Let $J \subset \mathbb{R}$ be an interval whose infimum, $\inf J$, is zero.

Consider the first-order evolution equation

$$\begin{cases} u'(t) & = Au(t) + f(t), \ t > 0 \\ u(0) & = u_0 \end{cases} \tag{8.1}$$

where $A : D(A) \subset \mathbb{X} \mapsto \mathbb{X}$ is a sectorial linear operator whose associated analytic semi-group will be denoted $(T(t))_{t \geq 0}$ and $f : J \mapsto \mathbb{X}$ is a continuous function.

Our main objective in this subsection consists of studying the existence of solutions to Eq. (8.1) when J is either $[0, T]$ or $\mathbb{R}_+ = [0, \infty)$ where $T > 0$ is a constant.

In this chapter, various types of solutions will be discussed. We basically follow and adopt definitions from Lunardi [87, Definition 4.1.1, Pages 123-124].

Definition 8.1 Let $f : [0, T] \mapsto \mathbb{X}$ be a continuous function and let $u_0 \in \mathbb{X}$. A function $u \in C([0, T]; D(A)) \cap C^1([0, T]; \mathbb{X})$ that satisfies,

$$u'(t) = Au(t) + f(t) \ \text{ for each } \ t \in [0, T] \ \text{ and } \ u(0) = u_0,$$

is called a *strict solution* to Eq. (8.1) on the interval $J = [0, T]$.

© Springer Nature Switzerland AG 2018
T. Diagana, *Semilinear Evolution Equations and Their Applications*,
https://doi.org/10.1007/978-3-030-00449-1_8

Definition 8.2 Let $f : [0, T] \mapsto \mathbb{X}$ be a continuous function and let $u_0 \in \mathbb{X}$. A function $u \in C([0, T]; \mathbb{X})$ is called a *strong solution* to Eq. (8.1) on the interval $J = [0, T]$ if there exists a sequence $(u_n)_{n \in \mathbb{N}} \subset C([0, T]; D(A)) \cap C^1([0, T]; \mathbb{X})$ such that

$$\sup_{t \in [0,T]} \|u_n(t) - u(t)\| \to 0 \quad \text{and} \quad \sup_{t \in [0,T]} \|u'_n(t) - Au_n(t) - f(t)\| \to 0$$

as $n \to \infty$.

Definition 8.3 Let $f : (0, T] \mapsto \mathbb{X}$ be a continuous function. Any function $u \in C([0, T]; \mathbb{X}) \cap C((0, T]; D(A)) \cap C^1((0, T]; \mathbb{X})$ that satisfies,

$$u'(t) = Au(t) + f(t) \quad \text{for each } t \in (0, T] \text{ and } u(0) = u_0,$$

is called a *classical solution* to Eq. (8.1) on the interval $J = [0, T]$.

Definition 8.4 Let $f : [0, \infty) \mapsto \mathbb{X}$ be a continuous function. A function $u : [0, \infty) \mapsto \mathbb{X}$ is said to be a strict (respectively, classical or strong) solution to Eq. (8.1) on the interval $J = [0, \infty)$, if for every $T > 0$, the restriction of the function u to the interval $[0, T]$ is a strict (respectively, classical or strong) solution to Eq. (8.1) on the interval $[0, T]$.

Definition 8.5 Let $f \in L^1((0, T); \mathbb{X})$ and let $u_0 \in \mathbb{X}$. A function u is called a *mild solution* to Eq. (8.1) if it can be written as follows,

$$u(t) = T(t)u_0 + \int_0^t T(t - s)f(s)ds, \quad t \in [0, T]. \tag{8.2}$$

Let us make a few remarks upon the notions which we have just introduced.

Remark 8.6

i) If u is a strict solution to Eq. (8.1), then the function u satisfies,

$$u'(t) = Au(t) + f(t) \quad \text{for each } t \in [0, T] \text{ and } u(0) = u_0.$$

Consequently,

$$u'(0) = Au(0) + f(0) = Au_0 + f(0)$$

which yields two things. First, the previous equation makes sense only if u_0 belongs to $D(A)$. Second, $u'(0) = Au_0 + f(0)$ must belong to $\overline{D(A)}$.

ii) If u is a classical solution, then one must have $u_0 \in \overline{D(A)}$. In this event, if in addition, $f \in L^1((0, T); \mathbb{X}) \cap C((0, T]; \mathbb{X})$, then

$$u(t) = T(t)u_0 + \int_0^t T(t - s)f(s)ds, \quad t \in [0, T].$$

iii) If u is a mild solution to Eq. (8.1), then using the fact that A is a sectorial linear operator [see ii) of Proposition 3.12], it follows that there exists a constant $C > 0$ such that u satisfies the following estimate,

$$\|u\| \leq C\left(\|u_0\| + \int_0^t \|f(s)\|ds\right) \text{ for all } t \in [0, T].$$

Theorem 8.7 ([88, Lemma 4.2.5, Pages 45 and 46]) *Let* $f : (0, T] \mapsto \mathbb{X}$ *be a bounded continuous function and let* $u_0 \in \overline{D(A)}$. *If* u *is a mild solution to Eq. (8.1), then the following statements are equivalent.*

i) $u \in C((0, T]; D(A))$;
ii) $u \in C^1((0, T]; \mathbb{X})$;
iii) u *is a classical solution to Eq. (8.1).*

If $f \in C([0, T]; \mathbb{X})$, *then the following statements are equivalent:*

i) $u \in C([0, T]; D(A))$;
ii) $u \in C^1([0, T]; \mathbb{X})$;
iii) u *is a strict solution to Eq. (8.1).*

8.1.2 Existence of Almost Periodic Solutions

The main objective here consists of studying the existence of bounded and almost periodic solutions to first-order evolution equations in the case when the analytic semi-group $(T(t))_{t \geq 0}$ associated with our sectorial operator $A : D(A) \subset \mathbb{X} \mapsto \mathbb{X}$ is hyperbolic, that is,

$$\sigma(A) \cap i\mathbb{R} = \{\emptyset\}. \tag{8.3}$$

From Proposition 3.12 it follows that there exist constants $M_0, M_1 > 0$ such that

$$\|T(t)\| \leq M_0 e^{\omega t}, \quad t > 0, \tag{8.4}$$

$$\|t(A - \omega)T(t)\| \leq M_1 e^{\omega t}, \quad t > 0. \tag{8.5}$$

Since the semi-group $(T(t))_{t \geq 0}$ is assumed to be hyperbolic, then there exists a projection P and constants $M, \delta > 0$ such that $T(t)$ commutes with P, $N(P)$ is invariant with respect to $T(t)$, $T(t) : R(Q) \mapsto R(Q)$ is invertible, and the following hold

$$\|T(t)Px\| \leq Me^{-\delta t}\|x\| \qquad \text{for } t \geq 0, \tag{8.6}$$

$$\|T(t)Qx\| \leq Me^{\delta t}\|x\| \qquad \text{for } t \leq 0, \tag{8.7}$$

where $Q := I - P$ and, for $t \leq 0$, $T(t) := (T(-t))^{-1}$.

Definition 8.8 Let $\alpha \in (0, 1)$. A Banach space $(\mathbb{X}_\alpha, \| \cdot \|_\alpha)$ is said to be an intermediate space between $D(A)$ and $\mathbb{X}$, or a space of class $\mathscr{J}_\alpha$, if $D(A) \subset \mathbb{X}_\alpha \subset \mathbb{X}$ and there is a constant $c > 0$ such that

$$\|x\|_\alpha \leq c\|x\|^{1-\alpha}\|x\|_A^\alpha, \qquad x \in D(A), \tag{8.8}$$

where $\| \cdot \|_A$ is the graph norm of A.

Precise examples of the intermediate space $\mathbb{X}_\alpha$ include $D((-A^\alpha))$ for $\alpha \in (0, 1)$, the domains of the fractional powers of A, the real interpolation spaces $D_A(\alpha, \infty)$, $\alpha \in (0, 1)$, defined as the space of all $x \in \mathbb{X}$ such

$$[x]_\alpha = \sup_{0<t\leq 1} \|t^{1-\alpha} AT(t)x\| < \infty.$$

with the norm

$$\|x\|_\alpha = \|x\| + [x]_\alpha,$$

the abstract Hölder spaces $D_A(\alpha) := \overline{D(A)}^{\|\cdot\|_\alpha}$ as well as complex interpolation spaces $[\mathbb{X}, D(A)]_\alpha$.

For a given hyperbolic analytic semi-group $(T(t))_{t\geq 0}$, it can be checked that similar estimations as both Eqs. (8.6) and (8.7) still hold with the α-norms $\| \cdot \|_\alpha$. In fact, as the part of A in $R(Q)$ is bounded, it follows from Eq. (8.7) that

$$\|AT(t)Qx\| \leq C'e^{\delta t}\|x\| \text{ for } t \leq 0.$$

Thus from Eq. (8.8) there exists a constant $c(\alpha) > 0$ such that

$$\|T(t)Qx\|_\alpha \leq c(\alpha)e^{\delta t}\|x\| \quad \text{for } t \leq 0. \tag{8.9}$$

In addition to the above, the following holds

$$\|T(t)Px\|_\alpha \leq \|T(1)\|_{B(\mathbb{X},\mathbb{X}_\alpha)}\|T(t-1)Px\|, \quad t \geq 1,$$

and hence from Eq. (8.6), one obtains

$$\|T(t)Px\|_\alpha \leq M'e^{-\delta t}\|x\|, \qquad t \geq 1,$$

where M' depends on α. For $t \in (0, 1]$, by Eqs. (8.5) and (8.8),

$$\|T(t)Px\|_\alpha \leq M''t^{-\alpha}\|x\|.$$

Hence, there exist constants $M(\alpha) > 0$ and $\gamma > 0$ such that

$$\|T(t)Px\|_\alpha \leq M(\alpha)t^{-\alpha}e^{-\gamma t}\|x\| \qquad \text{for } t > 0. \tag{8.10}$$

Consider the differential equation

$$u'(t) = Au(t) + f(t), \quad t \in \mathbb{R} \tag{8.11}$$

where $A : D(A) \subset \mathbb{X} \mapsto \mathbb{X}$ is a sectorial linear operator for which Eq. (8.3) holds and $f : \mathbb{R} \mapsto \mathbb{X}$ is a bounded continuous function.

Definition 8.9 A function $u \in BC(\mathbb{R}, \mathbb{X})$ is called a *mild solution* to Eq. (8.11) on $\mathbb{R}$ if for all $\tau \in \mathbb{R}$,

$$u(t) = T(t - \tau)u(\tau) + \int_{\tau}^{t} T(t - s)f(s)ds, \quad t \geq \tau. \tag{8.12}$$

Proposition 8.10 *If $f \in BC(\mathbb{R}, \mathbb{X})$, then Eq. (8.11) has a unique mild solution $u \in BC(\mathbb{R}, \mathbb{X})$ given by*

$$u(t) = \int_{-\infty}^{t} T(t - s)Pf(s)ds - \int_{t}^{\infty} T(t - s)(I - P)f(s)ds, \quad t \in \mathbb{R}. \tag{8.13}$$

Moreover, if $f \in C^{0,\alpha}(\mathbb{R}, \mathbb{X})$ for some $\alpha \in (0, 1)$, then u given above is a strict solution to Eq. (8.11) that belongs to $C^{0,\alpha}(\mathbb{R}, D(A))$.

Proof Clearly, the function given in Eq. (8.13), that is,

$$u(t) = \int_{-\infty}^{t} T(t - s)Pf(s)ds - \int_{t}^{\infty} T(t - s)(I - P)f(s)ds, \quad t \in \mathbb{R}$$

is well defined and satisfies

$$u(t) = T(t - s)u(s) + \int_{s}^{t} T(t - s)f(s)ds, \quad \text{for all } t, s \in \mathbb{R}, \ t \geq s. \tag{8.14}$$

Consequently, u is a mild solution to Eq. (8.11).

For the uniqueness, let v be another mild solution to Eq. (8.11). Thus using the projections P and $Q = I - P$, one obtains

$$Pv(t) = T(t - s)Pv(s) + \int_{s}^{t} T(t - s)Pf(s)ds, \quad \text{for all } t, s \in \mathbb{R}, \ t \geq s, \tag{8.15}$$

and

$$Qv(t) = T(t - s)Qv(s) + \int_{s}^{t} T(t - s)Qf(s)ds, \quad \text{for all } t, s \in \mathbb{R}, \ t \geq s. \tag{8.16}$$

Using the fact that v is bounded and Eqs. (8.9)–(8.10), letting $s \to -\infty$ in Eq. (8.15) (respectively, letting $s \to \infty$ in Eq. (8.16)), we obtain

$$Pv(t) = \int_{-\infty}^{t} T(t - s)Pf(s)ds, \quad \text{for all } t \in \mathbb{R},$$

and

$$Qv(t) = -\int_t^\infty T(t-s)Qf(s)ds, \quad \text{for all } t \in \mathbb{R},$$

which yields

$$v(t) = Pv(t) + Qv(t) = \int_{-\infty}^t T(t-s)Pf(s)ds$$

$$-\int_t^\infty T(t-s)Qf(s)ds = u(t), \quad \forall t \in \mathbb{R}.$$

Therefore, $u = v$.

Using [88, Lemma 3.3.1 and Lemma 3.3.3], it can be shown that if $f \in C^{0,\alpha}(\mathbb{R}, \mathbb{X})$ for some $\alpha \in (0, 1)$, then u belongs to $C^{0,\alpha}(\mathbb{R}, D(A))$.

We have

Corollary 8.11 *If $f \in AP(\mathbb{X})$, then Eq. (8.11) has a unique mild solution $u \in AP(\mathbb{X})$ given by*

$$u(t) = \int_{-\infty}^t T(t-s)Pf(s)ds - \int_t^\infty T(t-s)(I-P)f(s)ds, \quad t \in \mathbb{R}. \quad (8.17)$$

In particular, if f is continuous and T-periodic, then the mild solution u is also T-periodic.

Proof Using the fact that $AP(\mathbb{X}) \subset BC(\mathbb{R}, \mathbb{X})$, it follows that Eq. (8.11) has a unique mild solution $u \in BC(\mathbb{R}, \mathbb{X})$ given by

$$u(t) = \int_{-\infty}^t T(t-s)Pf(s)ds - \int_t^\infty T(t-s)(I-P)f(s)ds, \quad t \in \mathbb{R}.$$

To complete the proof, we have to show that $u \in AP(\mathbb{X})$. Since $f \in AP(\mathbb{X})$, for all $\varepsilon > 0$, there exists $\ell(\varepsilon) > 0$ such that for all $a \in \mathbb{R}$, the interval $(a, a + \ell(\varepsilon))$ contains a τ such that

$$\|f(t+\tau) - f(t)\| < \varepsilon \quad (8.18)$$

for all $t \in \mathbb{R}$.

Now

$$u(t+\tau) - u(t) = \int_{-\infty}^0 T(-s)P\Big(f(s+t+\tau) - f(s+t)\Big)ds$$

$$+ \int_0^\infty T(-s)Q\Big((s+t+\tau) - f(s+t)\Big)ds$$

which, for $\alpha \in (0, 1)$, yields

$$
\begin{aligned}
\|u(t + \tau) - u(t)\|_\alpha &\leq \int_{-\infty}^{0} \|T(-s)P\Big(f(s + t + \tau) - f(s + t)\Big)\|_\alpha ds \\
&\quad + \int_{0}^{\infty} \|T(-s)Q\Big((s + t + \tau) - f(s + t)\Big)\|_\alpha ds \\
&\leq c(\alpha) \int_{-\infty}^{0} e^{\delta s} \|f(s - t - \tau) - f(s - t)\| ds \\
&\quad + M(\alpha) \int_{0}^{\infty} e^{-\gamma s} s^{-\alpha} \|f(s - t - \tau) - f(s - t)\| ds
\end{aligned}
$$

by using Eqs. (8.9)–(8.10).

To conclude, one makes use of both Eq. (8.18) and the Lebesgue dominated convergence theorem.

8.2 Semilinear First-Order Evolution Equations

8.2.1 Existence of Mild and Classical Solutions

Consider the first-order semilinear evolution equation

$$
\begin{cases}
u'(t) = Au(t) + F(t, u(t)), & t > 0 \\
u(0) = u_0
\end{cases}
\tag{8.19}
$$

where $A : D(A) \subset \mathbb{X} \mapsto \mathbb{X}$ is a sectorial linear operator whose corresponding analytic semi-group is $(T(t))_{t \geq 0}$ and $F : [0, T] \times \mathbb{X} \mapsto \mathbb{X}$ is a jointly continuous function and locally Lipschitz in the second variable, that is, there exist $R > 0$ and $L > 0$ such that

$$
\|F(t, u) - F(t, v)\| \leq L \|u - v\|
\tag{8.20}
$$

for all $t \in [0, T]$ and $u, v \in B(0, R)$.

Let $J = [0, T_0)$ or $[0, T_0]$ where $T_0 \leq T$. As in the linear case, we have the following definitions for classical, strict, and mild solutions.

Definition 8.12 ([88]) A function $u : J \mapsto \mathbb{X}$ is said to be a strict solution to Eq. (8.19) in J, if u is continuous with values in $D(A)$ and differentiable with values in $\mathbb{X}$ in the interval J, and satisfies Eq. (8.19).

Definition 8.13 ([88]) A function $u : J \mapsto \mathbb{X}$ is said to be a classical solution to Eq. (8.19) in J, if u is continuous with values in $D(A)$ and differentiable with values in $\mathbb{X}$ in the interval $J \setminus \{0\}$, continuous in the interval J with values in $D(A)$, and satisfies Eq. (8.19).

Definition 8.14 ([88]) A function $u : J \mapsto \mathbb{X}$ is called a *mild solution* to Eq. (8.19) if it is continuous with values in $D(A)$ in the interval $J \setminus \{0\}$, and satisfies

$$u(t) = T(t)u_0 + \int_0^t T(t-s)F(s, u(s))ds, \quad t \in J. \tag{8.21}$$

Under some suitable conditions (see [87]) it can be shown that every mild solution to Eq. (8.19) is a classical (or strict) solution.

Theorem 8.15 ([88]) *Suppose $F : [0, T] \times \mathbb{X} \mapsto \mathbb{X}$ is jointly continuous and satisfies Eq. (8.20). Then for every $v \in \mathbb{X}$ there exist constants $r, \delta > 0$, $K > 0$ such that for $\|u_0 - v\| \leq r$, then Eq. (8.19) has a unique mild solution $u = u(\cdot, u_0) \in BC((0, \delta]; \mathbb{X})$. The mild solution u belongs to $C([0, \delta]; \mathbb{X})$ if and only if $u_0 \in \overline{D(A)}$. Further, for $u_0, u_1 \in B(v, r)$, the following holds,*

$$\|u(t, u_0) - u(t, u_1)\| \leq K \|u_0 - u_1\|, \quad t \in [0, \delta].$$

8.2.2   Existence Results on the Real Number Line

Consider the differential equation

$$u'(t) = Au(t) + F(t, u(t)), \quad t \in \mathbb{R} \tag{8.22}$$

where $A : D(A) \subset \mathbb{X} \mapsto \mathbb{X}$ is a sectorial linear operator for which Eq. (8.3) holds and $F : \mathbb{R} \times \mathbb{X}_\alpha \mapsto \mathbb{X}$ for some $\alpha \in (0, 1)$ is a jointly continuous function and globally Lipschitz in the second variable, that is, there exists a constant $L > 0$ such that

$$\|F(t, u) - F(t, v)\| \leq L \|u - v\|_\alpha \tag{8.23}$$

for all $t \in \mathbb{R}$ and $u, v \in \mathbb{X}_\alpha$.

Definition 8.16 A mild solution to Eq. (8.22) is any function $u : \mathbb{R} \mapsto \mathbb{X}_\alpha$ which satisfies the following variation of constants formula,

$$u(t) = T(t-s)u(s) + \int_s^t T(t-\sigma)F(\sigma, u(\sigma))d\sigma \tag{8.24}$$

for all $t \geq s$, $t, s \in \mathbb{R}$.

Using Corollary 8.11 in which we let $f(t) := F(t, u(t))$ and under some additional assumptions, we obtain the next theorem.

Theorem 8.17 *Under Eqs. (8.3) and (8.23), if $F \in AP(\mathbb{R} \times \mathbb{X}_\alpha, \mathbb{X})$, then Eq. (8.22) has a unique almost periodic mild solution if L is small enough.*

Remark 8.18 A generalization of Theorem 8.17 to the case when the linear operator A is replaced with $A(t)$ is given by Theorem 8.20.

8.3 Nonautonomous First-Order Evolution Equations

8.3.1 Existence of Almost Periodic Mild Solutions

Consider the nonautonomous differential equation

$$u'(t) = A(t)u(t) + f(t, u(t)) \tag{8.25}$$

where $A(t)$ for $t \in \mathbb{R}$ be a family of linear operators on $\mathbb{X}$ whose domains $D(A(t)) = D$ are constant for all $t \in \mathbb{R}$ and the function $f : \mathbb{R} \times \mathbb{X}_\alpha \mapsto \mathbb{X}$ is continuous and globally lipschitzian, that is, there is $k > 0$ such that

$$\| f(t, x) - f(t, y) \| \le k \, \| x - y \|_\alpha \quad \text{for all } t \in \mathbb{R} \text{ and } x, y \in \mathbb{X}_\alpha. \tag{8.26}$$

To study the almost periodicity of the solutions of Eq. (8.25), we assume that the following holds:

(H.820) The family of linear operators $A(t)$ satisfy the Aquistapace–Terreni conditions.

(H.821) The evolution family U generated by $A(\cdot)$ has an exponential dichotomy with constants $N, \delta > 0$ and dichotomy projections $P(t)$ for $t \in \mathbb{R}$.

(H.822) There exists $0 \le \alpha < \beta < 1$ such that

$$\mathbb{X}_\alpha^t = \mathbb{X}_\alpha \quad \text{and} \quad \mathbb{X}_\beta^t = \mathbb{X}_\beta$$

for all $t \in \mathbb{R}$, with uniform equivalent norms.

(H.823) $R(\omega, A(\cdot)) \in AP(\mathbb{R}, B(\mathbb{X}))$ with pseudo periods $\tau = \tau_\epsilon$ belonging to sets $\mathscr{P}(\epsilon, A)$.

Definition 8.19 By a mild solution Eq. (8.25) we mean every continuous function $x : \mathbb{R} \mapsto \mathbb{X}_\alpha$, which satisfies the following variation of constants formula

$$x(t) = U(t, s)x(s) + \int_s^t U(t, \sigma)f(\sigma, x(\sigma))d\sigma \quad \text{for all } t \ge s, \ t, s \in \mathbb{R}. \tag{8.27}$$

In order to study the existence of almost periodic mild solution to the semilinear evolution equation Eq. (8.25), we first study the existence of almost periodic mild solution to the inhomogeneous evolution equation

$$x'(t) = A(t)x(t) + g(t), \quad t \in \mathbb{R}. \tag{8.28}$$

We have

Theorem 8.20 *Suppose that assumptions (H.820)–(H.823) hold. If $g \in BC(\mathbb{R}, \mathbb{X})$, then*

(i) *Equation (8.28) has a unique bounded mild solution $x : \mathbb{R} \mapsto \mathbb{X}_\alpha$ given by*

$$x(t) = \int_{-\infty}^{t} U(t, s) P(s) g(s) ds - \int_{t}^{+\infty} \tilde{U}(t, s) Q(s) g(s) ds. \qquad (8.29)$$

(ii) *If $g \in AP(\mathbb{R}, \mathbb{X})$, then $x \in AP(\mathbb{R}, \mathbb{X}_\alpha)$.*

Proof

(i) Since g is bounded, we know from [37] that the function x given by (8.29) is the unique bounded mild solution to Eq. (8.28). To prove that x is bounded in $\mathbb{X}_\alpha$, we make use of Proposition 3.27 to obtain,

$$\|x(t)\|_\alpha \le c \, \|x(t)\|_\beta$$

$$\le c \int_{-\infty}^{t} \|U(t, s) P(s) g(s)\|_\beta \, ds + c \int_{t}^{+\infty} \|\tilde{U}(t, s) Q(s) g(s)\|_\beta \, ds$$

$$\le cc(\beta) \int_{-\infty}^{t} e^{-\frac{\delta}{2}(t-s)} (t - s)^{-\beta} \|g(s)\| \, ds$$

$$+ cm(\beta) \int_{t}^{+\infty} e^{-\delta(s-t)} \|g(s)\| \, ds$$

$$\le cc(\beta) \|g\|_\infty \int_{0}^{+\infty} e^{-\sigma} \left(\frac{2\sigma}{\delta}\right)^{-\beta} \frac{2d\sigma}{\delta} + cm(\beta) \|g\|_\infty \int_{0}^{+\infty} e^{-\delta\sigma} d\sigma$$

$$\le cc(\beta) \delta^\alpha \Gamma(1 - \beta) \|g\|_\infty + cm(\beta) \delta^{-1} \|g\|_\infty,$$

and hence

$$\|x(t)\|_\alpha \le c \, \|x(t)\|_\beta \le c[c(\beta) \delta^\beta \Gamma(1 - \beta) + m(\beta) \delta^{-1}] \|g\|_\infty. \qquad (8.30)$$

(ii) Let $\epsilon > 0$ and $\mathscr{P}(\epsilon, A, f)$ be the set of pseudo periods for the almost periodic function $t \mapsto (f(t), R(\omega, A(t)))$. We know, from [89, Theorem 4.5] that x, as an $\mathbb{X}$-valued function is almost periodic. Hence, there exists a number $\tau \in \mathscr{P}((\frac{\varepsilon}{c'})^{\frac{\beta}{\beta-\alpha}}, A, f)$ such that

$$\|x(t + \tau) - x(t)\| \le \left(\frac{\varepsilon}{c'}\right)^{\frac{\beta}{\beta-\alpha}} \qquad \text{for all } t \in \mathbb{R}.$$

For $\theta = \frac{\alpha}{\beta}$, the reiteration theorem implies that $\mathbb{X}_\alpha = (\mathbb{X}, \mathbb{X}_\beta)_{\theta,\infty}$. Using the property of interpolation and Eq. (8.30), we obtain

$$\|x(t+\tau) - x(t)\|_\alpha \le c(\alpha, \beta)\|x(t+\tau) - x(t)\|^{\frac{\beta-\alpha}{\beta}} \|x(t+\tau) - x(t)\|_\beta^{\frac{\alpha}{\beta}}$$

$$\le c(\alpha, \beta)2^{\frac{\alpha}{\beta}} \left(c[c(\beta)\delta^\beta \Gamma(1-\beta) + m(\beta)\delta^{-1}]\|g\|_\infty\right)^{\frac{\alpha}{\beta}}$$

$$\|x(t+\tau) - x(t)\|^{\frac{\beta-\alpha}{\beta}}$$

$$:= c'\|x(t+\tau) - x(t)\|^{\frac{\beta-\alpha}{\beta}},$$

and hence

$$\|x(t+\tau) - x(t)\|_\alpha \le \varepsilon$$

for $t \in \mathbb{R}$.

To show the existence of almost periodic solutions to Eq. (8.25), let $y \in AP(\mathbb{R}, \mathbb{X}_\alpha)$ and $f \in AP(\mathbb{R} \times \mathbb{X}_\alpha, \mathbb{X})$. Using the theorem of composition of almost periodic functions (Theorem 4.18) we deduce that the function $g(\cdot) := f(\cdot, y(\cdot)) \in AP(\mathbb{R}, \mathbb{X})$, and from Theorem 8.20, the semilinear equation (Eq. (8.25)) has a unique mild solution $x \in AP(\mathbb{R}, \mathbb{X}_\alpha)$ given by

$$x(t) = \int_{-\infty}^t U(t,s)P(s)f(s, y(s))ds - \int_t^{+\infty} \widetilde{U}(t,s)Q(s)f(s, y(s))ds, \quad t \in \mathbb{R}.$$

Define the nonlinear operator $F : AP(\mathbb{R}, \mathbb{X}_\alpha) \mapsto AP(\mathbb{R}, \mathbb{X}_\alpha)$ by

$$(Fy)(t) := \int_{-\infty}^t U(t,s)P(s)f(s, y(s))ds$$

$$- \int_t^{+\infty} \widetilde{U}(t,s)Q(s)f(s, y(s))ds, \quad t \in \mathbb{R}.$$

Now for any $x, y \in AP(\mathbb{R}, \mathbb{X}_\alpha)$,

$$\|Fx(t) - Fy(t)\|_\alpha \le c(\alpha) \int_{-\infty}^t e^{-\delta(t-s)}(t-s)^{-\alpha} \|f(s, y(s)) - f(s, x(s))\| ds$$

$$+ c(\alpha) \int_t^{+\infty} e^{-\delta(t-s)} \|f(s, y(s)) - f(s, x(s))\| ds.$$

$$\le k[c(\alpha)\delta^{-\alpha}\Gamma(1-\alpha) + m(\alpha)\delta^{-1}] \|x - y\|_\infty \quad \text{for all } t \in \mathbb{R}.$$

By taking k small enough, that is, $k < (c(\alpha)\delta^{\alpha}\Gamma(1-\alpha) + m(\alpha)\delta^{-1})^{-1}$, the operator F becomes a contraction on $AP(\mathbb{R}, \mathbb{X}_{\alpha})$ and hence has a unique fixed point in $AP(\mathbb{R}, \mathbb{X}_{\alpha})$, which obviously is the unique $\mathbb{X}_{\alpha}$-valued almost periodic solution to Eq. (8.25).

The previous discussion can be formulated as follows:

Theorem 8.21 *Let* $\alpha \in (0, 1)$. *Suppose that assumptions (H.820)–(H.821)–(H.822)–(H.823) hold and that* $f \in AP(\mathbb{R} \times \mathbb{X}_{\alpha}, \mathbb{X})$ *with* $k < (c(\alpha)\delta^{-\alpha}\Gamma(1-\alpha) + m(\alpha)\delta^{-1})^{-1}$. *Then Eq. (8.25) has a unique mild solution* x *in* $AP(\mathbb{R}, \mathbb{X}_{\alpha})$.

8.4   Exercises

1. Prove Theorem 8.7
2. Prove Theorem 8.15.
3. Prove Theorem 8.17

8.5   Comments

The preliminary results of this chapter are taken from Lunardi [87, 88]. Let us point out that the setting of Sect. 8.1.2 follows that of Boulite et al. [30]. The proofs of Theorem 8.20 and Theorem 8.21 discussed follow Baroun et al. [20]. The existence of mild solutions for similar evolution equations can be obtained in the cases when the operator A is not necessarily sectorial. For these cases, we refer the interested readers to Pazy [100] and Engel and Nagel [55]. The existence results obtained when the forcing term is almost periodic can be extended to more general classes of functions including almost automorphic and pseudo-almost periodic or pseudo-almost automorphic functions, see, e.g., Diagana [47].

Chapter 9
Semilinear Fractional Evolution Equations

9.1 Introduction

Let $\alpha \in (0, 1]$. The main objective of this chapter consists of acquainting the reader with the fast-growing theory of fractional evolution equations. More precisely, we study sufficient conditions for the existence of classical (respectively, mild) solutions for the inhomogeneous fractional Cauchy problem

$$\begin{cases} \mathbb{D}_t^\alpha u(t) = Au(t) + f(t), & t > 0 \\ u(0) = u_0 \in \mathbb{X} \end{cases}$$

and its corresponding semilinear evolution equation

$$\begin{cases} \mathbb{D}_t^\alpha u(t) + Au(t) = F(t, u(t)), & t > 0, \\ u(0) = u_0 \in \mathbb{X}, \end{cases}$$

where $\mathbb{D}_t^\alpha$ is the fractional derivative of order α in the sense of Caputo, $A : D(A) \subset \mathbb{X} \mapsto \mathbb{X}$ is a closed linear operator on a complex Banach space $\mathbb{X}$ (respectively, $A \in \Sigma_\omega^\gamma(\mathbb{X})$ where $\gamma \in (-1, 0)$ and $0 < \omega < \frac{\pi}{2}$), and $f : [0, \infty) \mapsto \mathbb{X}$ and $F : [0, \infty) \times \mathbb{X} \mapsto \mathbb{X}$ are continuous functions satisfying some additional conditions. Under some appropriate assumptions, various existence results are discussed.

The main tools utilized to establish the existence of classical (respectively, mild) solutions to the above-mentioned fractional evolutions are the so-called $(\alpha, \alpha)^\beta$-resolvent families $S_\alpha^\beta(\cdot)$ and almost sectorial operators. Additional details on these classes of operators can be found in Keyantuo et al. [75] and Wang et al. [110]. For additional readings upon the topics discussed in this chapter, we refer to [6, 21–23, 37, 42], etc.

© Springer Nature Switzerland AG 2018

T. Diagana, *Semilinear Evolution Equations and Their Applications*,
https://doi.org/10.1007/978-3-030-00449-1_9

9.2   Fractional Calculus

Let $(\mathbb{X}, \|\cdot\|)$ be a complex Banach space. If $J \subset \mathbb{R}$ is an interval and if $(V, \|\cdot\|) \subset \mathbb{X}$ is a (normed) subspace, then $C(J; V)$ (respectively, $C^{(k)}(I; V)$ for $k \in \mathbb{N}$) will denote the collection of all continuous functions from J into V (respectively, the collection of all functions of class C^k which go from J into V).

Definition 9.1 If $f : \mathbb{R}_+ \mapsto \mathbb{R}$ and $g : \mathbb{R}_+ \mapsto \mathbb{X}$ are functions, we define their convolution, if it exists, as follows:

$$(f * g)(t) := \int_0^t f(t-s)g(s)ds, \quad t \geq 0.$$

Definition 9.2 If $u : \mathbb{R}_+ \mapsto \mathbb{X}$ is a function, then its Riemann–Liouville fractional derivative of order β is defined by

$$D_t^\beta u(t) := \frac{d^n}{dt^n}\left[\int_0^t g_{n-\beta}(t-s)u(s)ds\right], \quad t > 0$$

where $n := \lceil \beta \rceil$ is the smallest integer greatest than or equal to β, and

$$g_\beta(t) := \frac{t^{\beta-1}}{\Gamma(\beta)}, \quad t > 0, \ \beta > 0,$$

with $g_0 = \delta_0$ (the Dirac measure concentrated at 0).

Note that $g_{\alpha+\beta} = g_\alpha * g_\beta$ for all $\alpha, \beta \geq 0$.

Definition 9.3 The Caputo fractional derivative of order $\beta > 0$ of a function $u :$ $\mathbb{R}_+ \mapsto \mathbb{X}$ is defined by

$$\mathbb{D}_t^\beta u(t) := D_t^{n-\beta} u^{(n)}(t) = \int_0^t g_{n-\beta}(t-s)u^{(n)}(s)ds,$$

where $n := \lceil \beta \rceil$.

We have the following additional relationship between Riemann–Liouville and Caputo fractional derivatives:

$$\mathbb{D}_t^\beta f(t) = D_t^\beta\left(f(t) - \sum_{k=0}^{n-1} f^{(k)}(0)g_{k+1}(t)\right), \quad t > 0,$$

where $n := \lceil \beta \rceil$.

Definition 9.4 If $f : \mathbb{R}_+ \mapsto \mathbb{X}$ is integrable, then its Laplace transform is defined by

$$(\mathbb{L}f)(z) = \widehat{f}(z) := \int_0^\infty e^{-zt} f(t)dt$$

provided this integral converges absolutely for some $z \in \mathbb{C}$.

Among other things, if $\lceil \alpha \rceil = n \geq 1$, then

$$\widehat{D_t^\alpha f}(z) = z^\alpha \widehat{f}(z) - \sum_{k=0}^{n-1} z^{\alpha-k-1} f^{(k)}(0)$$

and

$$\widehat{D_t^\alpha f}(z) = z^\alpha \widehat{f}(z) - \sum_{k=0}^{n-1} (g_{n-\alpha} * f)^{(k)}(0) z^{n-1-k}$$

where z^α is uniquely defined as $z^\alpha = |z|^\alpha e^{i \arg z}$ with $-\pi < \arg z < \pi$.

Let $k \in \mathbb{N}$. If $u \in C^{k-1}(\mathbb{R}_+; \mathbb{X})$ and $v \in C^k(\mathbb{R}_+; \mathbb{X})$, then for every $t \geq 0$,

$$\frac{d^k}{dt^k}[(u * v)(t)] = \sum_{j=0}^{k-1} u^{(k-1-j)}(t) v^{(j)}(0) + (u * v^{(k)})(t)$$

$$= \sum_{j=0}^{k-1} \frac{d^{k-1}}{dt^{k-1}}\left[(g_j * u)(t) v^{(j)}(0)\right] + (u * v^{(k)})(t). \tag{9.1}$$

Define the generalized Mittag–Leffler special function $E_{\alpha,\beta}$ by

$$E_{\alpha,\beta}(z) = \sum_{k=0}^\infty \frac{z^k}{\Gamma(\alpha k + \beta)}$$

$$= \frac{1}{2\pi i} \int_\Gamma \frac{\lambda^{\alpha-\beta} e^\lambda}{\lambda^\alpha - z} d\lambda, \quad \alpha, \beta > 0, z \in \mathbb{C},$$

where Γ is a contour which starts and ends at $-\infty$ and encircles the disc

$$|\lambda| \leq |z|^{1/\alpha}$$

counter-clockwise.

In what follows, we set

$$E_\alpha(z) := E_{\alpha,1}(z),$$

and

$$e_\alpha(z) := E_{\alpha,\alpha}(z).$$

9.3   Inhomogeneous Fractional Differential Equations

9.3.1   Introduction

Let $\alpha \in (0, 1]$. In this section, we study the existence of classical (respectively, mild) solutions for the inhomogeneous fractional Cauchy problem

$$\begin{cases} \mathbb{D}_t^\alpha u(t) = Au(t) + f(t) \\ u(0) = u_0 \in \mathbb{X} \end{cases} \tag{9.2}$$

where $\mathbb{D}_t^\alpha$ is the fractional derivative of order α in the sense of Caputo, $A : D(A) \subset \mathbb{X} \mapsto \mathbb{X}$ is a closed linear operator on a complex Banach space $\mathbb{X}$, and $f : \mathbb{R}_+ \mapsto \mathbb{X}$ is a function satisfying some additional conditions.

As it was pointed out in Keyantuo et al. [75], Caputo fractional derivative is more appropriate for equations of the form Eq. (9.2) than the Riemann–Liouville fractional derivative. Indeed, Caputo fractional derivative requires that the solution u of the above Cauchy problem be known at $t = 0$ while that of Riemann–Liouville requires that it be known in a right neighborhood of $t = 0$.

Recall that if $\alpha = 1$, then there are two situations that can be considered. If A is the infinitesimal generator of a strongly continuous semi-group, then semi-group techniques can be used to establish the existence of solutions to Eq. (9.2). Now, if A is not the infinitesimal generator of a strongly continuous semi-group, the concept of exponentially bounded β-times integrated semi-groups can be utilized to deal with existence of solutions to the above Cauchy problem. Similarly, if $\alpha \in (0, 1)$, a family of strongly continuous linear operators $S_\alpha : \mathbb{R}_+ \mapsto B(\mathbb{X})$ can be used to establish the existence of solutions to the above Cauchy problem. Unfortunately, the previous concept is inappropriate for some important practical problems, see details in Keyantuo et al. [75]. This in fact is one of the main reasons that led Keyantuo et al. to introduce the concept of $(\alpha, \alpha)^\beta$-resolvent families (respectively, $(\alpha, 1)^\beta$-resolvent families), which generalizes naturally all the above-mentioned cases. Such a new concept will play a central role in this section.

9.3.2   Basic Definitions

Definition 9.5 ([75]) Let $A : D(A) \subset \mathbb{X} \mapsto \mathbb{X}$ be a closed linear operator and let $\alpha \in (0, 1]$ and $\beta \geq 0$. The operator A is called an $(\alpha, \alpha)^\beta$-resolvent family if there exist $\omega \geq 0$, $M \geq 0$, and a family of strongly continuous functions $T_\alpha^\beta : [0, \infty) \mapsto B(\mathbb{X})$ (respectively, $T_\alpha^\beta : (0, \infty) \mapsto B(\mathbb{X})$ in the case when $\alpha(1 + \beta) < 1$) such that,

i) $\left\| (g_1 * T_\alpha^\beta)(t) \right\| \leq Me^{\omega t}$ for all $t > 0$;

ii) $\left\{\lambda^{\alpha} : \Re e\, \lambda > \omega\right\} \subset \rho(A)$; and

$$(\lambda^{\alpha} I - A)^{-1} u = \lambda^{\alpha\beta} \int_0^{\infty} e^{-\lambda t} T_{\alpha}^{\beta}(t) u \, dt, \quad \Re e\, \lambda > \omega, \quad u \in \mathbb{X}.$$

Definition 9.6 ([75]) Let $A : D(A) \subset \mathbb{X} \mapsto \mathbb{X}$ be a closed linear operator and let $\alpha \in (0, 1]$ and $\beta \geq 0$. The operator A is called an $(\alpha, 1)^{\beta}$-resolvent family generator if there exist $\omega \geq 0$, $M \geq 0$, and a family of strongly continuous functions $S_{\alpha}^{\beta} : \mathbb{R}_+ \mapsto B(\mathbb{X})$ such that,

i) $\left\|(g_1 * S_{\alpha}^{\beta})(t)\right\| \leq M e^{\omega t}$ for $t \geq 0$;

ii) $\left\{\lambda^{\alpha} : \Re e\, \lambda > \omega\right\} \subset \rho(A)$; and

$$\lambda^{\alpha-1}(\lambda^{\alpha} I - A)^{-1} u = \lambda^{\alpha\beta} \int_0^{\infty} e^{-\lambda t} S_{\alpha}^{\beta}(t) u \, dt, \quad \Re e\, \lambda > \omega, \quad u \in \mathbb{X}.$$

Remark 9.7 A family of strongly continuous functions $T_{\alpha}^{\beta}(t)$ that satisfies items i)–ii) of Definition 9.5 is called the $(\alpha, \alpha)^{\beta}$-resolvent family generated by the linear operator A. And there is uniqueness of the $(\alpha, \alpha)^{\beta}$-resolvent family (respectively, $(\alpha, 1)^{\beta}$-resolvent family) associated with a given operator A.

In fact, there is a relationship between these two new notions. It is not hard to show that if A generates an $(\alpha, \alpha)^{\beta}$-resolvent family T_{α}^{β}, then it generates an $(\alpha, 1)^{\beta}$-resolvent family S_{α}^{β} (see [75] for details) and that both T_{α}^{β} and S_{α}^{β} are linked through the following identity,

$$S_{\alpha}^{\beta}(t)x = (g_{1-\alpha} * T_{\alpha}^{\beta})(t)x, \quad t \geq 0, \quad x \in \mathbb{X}.$$

Let us now collect a few additional properties of both $(\alpha, 1)^{\beta}$- and $(\alpha, \alpha)^{\beta}$-resolvent families.

Proposition 9.8 ([75]) *Let $A : D(A) \subset \mathbb{X} \mapsto \mathbb{X}$ be a closed linear operator and let $\alpha \in (0, 1]$ and $\beta \geq 0$. If A generates an $(\alpha, 1)^{\beta}$-resolvent family S_{α}^{β}, then the following hold,*

i) $S_{\alpha}^{\beta}(t)(D(A)) \subset D(A)$ and

$$AS_{\alpha}^{\beta}(t)x = S_{\alpha}^{\beta}(t)Ax$$

for all $x \in D(A)$ and $t \geq 0$.

ii) For all $x \in D(A)$,

$$S_{\alpha}^{\beta}(t)x = g_{\alpha\beta+1}(t)x + \int_0^t g_{\alpha}(t - s)AS_{\alpha}^{\beta}(s)x\,ds, \quad t \geq 0.$$

iii) For all $x \in \mathbb{X}$, $(g_\alpha * S_\alpha^\beta)(t)x \in D(A)$,

$$S_\alpha^\beta(t)x = g_{\alpha\beta+1}(t)x + A \int_0^t g_\alpha(t-s)S_\alpha^\beta(s)xds, \quad t \geq 0.$$

iv) $S_\alpha^\beta(0) = g_{\alpha\beta+1}(0)$; $S_\alpha^\beta(0) = I$ if $\beta = 0$ and $S_\alpha^\beta(0) = 0$ if $\beta > 0$.

Proposition 9.9 ([75]) *Let $A : D(A) \subset \mathbb{X} \mapsto \mathbb{X}$ be a closed linear operator and let $\alpha \in (0, 1]$ and $\beta \geq 0$. If A generates an $(\alpha, \alpha)^\beta$-resolvent family T_α^β, then the following hold,*

i) $T_\alpha^\beta(t)(D(A)) \subset D(A)$ *and*

$$AT_\alpha^\beta(t)x = T_\alpha^\beta(t)Ax$$

for all $x \in D(A)$ and $t > 0$.
ii) *For all $x \in D(A)$,*

$$T_\alpha^\beta(t)x = g_{\alpha(\beta+1)}(t)x + \int_0^t g_\alpha(t-s)AT_\alpha^\beta(s)xds, \quad t \geq 0.$$

iii) *For all $x \in \mathbb{X}$, $(g_\alpha * T_\alpha^\beta)(t)x \in D(A)$,*

$$T_\alpha^\beta(t)x = g_{\alpha(\beta+1)}(t)x + A \int_0^t g_\alpha(t-s)T_\alpha^\beta(s)xds, \quad t > 0.$$

iv) *If $\beta > 0$, then for every $x \in \overline{D(A)}$,*

$$\frac{1}{\Gamma(\alpha(1+\beta))} \lim_{t \to 0} t^{1-\alpha(1+\beta)} T_\alpha^\beta(t)x = x$$

if $\alpha(1+\beta) < 1$; $T_\alpha^\beta(0)x = x$ if $\alpha(1+\beta) = 1$; and $T_\alpha^\beta(0)x = 0$ if $\alpha(1+\beta) > 1$.
v) *If $\alpha(1+\beta) > 1$, then all the above equalities occur for $t \geq 0$. .*

A strongly continuous function $h : [0, \infty) \mapsto \mathbb{X}$ is called exponentially bounded if there exist constants $M, \omega \geq 0$ such that

$$\|h(t)\| \leq Me^{\omega t}$$

for all $t > 0$. In particular, an $(\alpha, \alpha)^\beta$-resolvent family T_α^β (respectively, $(\alpha, 1)^\beta$-resolvent family S_α^β) is exponentially bounded, there exist constants $M, \omega \geq 0$ such that

$$\|T_\alpha^\beta(t)\| \leq Me^{\omega t}$$

for all $t > 0$ (respectively, there exist constants $M', \omega' \geq 0$ such that

$$\|S_\alpha^\beta(t)\| \leq M' e^{\omega' t}$$

for all $t \geq 0$).

For more on $(\alpha, \alpha)^\beta$-resolvent families (respectively, $(\alpha, 1)^\beta$-resolvent families), we refer the reader to [75].

9.3.3 Existence of Classical and Mild Solutions

Definition 9.10 ([75]) A continuous function A function $u : [0, \infty) \mapsto D(A)$ is said to be a classical solution to Eq. (9.2) if $g_{1-\alpha} * (u - u(0)) : [0, \infty) \mapsto \mathbb{X}$ is a continuous function and Eq. (9.2) holds.

Definition 9.11 ([75]) A continuous function $u : [0, \infty) \mapsto \mathbb{X}$ is said to be a mild solution to Eq. (9.2) if $(g_\alpha * u)(t) \in D(A)$ for all $t \geq 0$ and

$$u(t) = u_0 + A \int_0^t g_\alpha(t-s)u(s)ds + \int_0^t g_\alpha(t-s)f(s)ds, \quad t \geq 0. \quad (9.3)$$

Theorem 9.12 ([75]) *Let $\alpha \subset (0, 1]$ and $\beta \geq 0$ and set $n = \lceil \beta \rceil$ and $k = \lceil \alpha\beta \rceil$. Suppose that A is the generator of an $(\alpha, 1)^\beta$-resolvent family S_α^β. Then the following hold,*

i) *For every $f \in C^{(k+1)}(\mathbb{R}_+; \mathbb{X})$, $f^{(l)}(0) \in D(A^{n+1-l})$ for $l = 0, 1, \ldots, k$, $\mathbb{D}_t^{\alpha\beta} f$ is exponentially bounded and $u_0 \in D(A^{n+1})$, Eq. (9.2) has a unique classical solution given by*

$$u(t) = D_t^{\alpha\beta} S_\alpha^\beta(t)u_0 + D_t^{\alpha\beta} \mathbb{D}_t^{1-\alpha}(S_\alpha^\beta * f)(t), \quad t \geq 0. \quad (9.4)$$

ii) *For every $f \in C^{(k)}(\mathbb{R}_+; \mathbb{X})$, $f^{(l)}(0) \in D(A^{n-l})$ for $l = 0, 1, \ldots, k - 1$, $\mathbb{D}_t^{\alpha\beta} f$ is exponentially bounded and $u_0 \in D(A^n)$, Eq. (9.2) has a unique mild solution given by Eq. (9.4).*

Corollary 9.13 ([75]) *Let $\alpha \in (0, 1]$ and $\beta \geq 0$ and set $n = \lceil \beta \rceil$ and $k = \lceil \alpha\beta \rceil$. Suppose that A generates an $(\alpha, \alpha)^\beta$-resolvent family T_α^β. And let S_α^β be the $(\alpha, 1)^\beta$-resolvent family generated by A. Then the following hold:*

(a) *For every $f \in C^{(k+1)}(\mathbb{R}_+; \mathbb{X})$, $f^{(j)}(0) \in D(A^{n+1-j})$ for $j = 0, 1, \ldots, k$, $\mathbb{D}_t^{\alpha\beta} f$ is exponentially bounded, and for every $u_0 \in D(A^{n+1})$, the unique classical solution to Eq. (9.2) is given by*

$$u(t) = D_t^{\alpha\beta} \left[S_\alpha^\beta(t)u_0 + \int_0^t T_\alpha^\beta(t-s)f(s)ds \right], \quad t \geq 0. \quad (9.5)$$

*(b) For every $f \in C^{(k)}(\mathbb{R}_+; \mathbb{X})$, $f^{(j)}(0) \in D(A^{n-j})$ for $j = 0, 1, \ldots, k-1$, $\mathbb{D}_t^{\alpha\beta} f$
is exponentially bounded and for every $u_0 \in D(A^n)$, the unique mild solution
to Eq. (9.2) is given by Eq. (9.5).*

9.4 Semilinear Fractional Differential Equations

9.4.1 Preliminaries and Notations

Let $\gamma \in (-1, 0)$ and let S_μ^0 (with $0 < \mu < \pi$) be the open sector defined by

$$\{z \in \mathbb{C} \setminus \{0\} : |\arg z| < \mu\}$$

and let S_μ be its closure, that is,

$$S_\mu := \{z \in \mathbb{C} \setminus \{0\} : |\arg z| \le \mu\} \cup \{0\}.$$

Definition 9.14 ([110]) Let $\gamma \in (-1, 0)$ and let $0 < \omega < \pi/2$. The set $\Sigma_\omega^\gamma(\mathbb{X})$
stands for the collection of all closed linear operators $A : D(A) \subset \mathbb{X} \to \mathbb{X}$
satisfying

i) $\sigma(A) \subset S_\omega$; and
ii) for every $\omega < \mu < \pi$ there exists a constant C_μ such that

$$\|(zI - A)^{-1}\| \le C_\mu |z|^\gamma \tag{9.6}$$

for all $z \in \mathbb{C} \setminus S_\mu$.

Definition 9.15 A linear operator $A : D(A) \subset \mathbb{X} \mapsto \mathbb{X}$ that belongs to $\Sigma_\omega^\gamma(\mathbb{X})$ will
be called an almost sectorial operator on $\mathbb{X}$.

Among other things, recall that if $A \in \Sigma_\omega^\gamma(\mathbb{X})$, then $0 \in \rho(A)$. Further, there
exist almost sectorial operators which are not sectorial, see, e.g., [109]. There are
many examples of almost sectorial operators in the literature, see, e.g., Wang et al.
[110].

Let $\alpha \in (0, 1)$. Our main objective in this section consists of studying the
existence of solutions to the following semilinear fractional differential equations

$$\begin{cases} \mathbb{D}_t^\alpha u(t) + Au(t) = f(t, u(t)), & t > 0, \\ u(0) = u_0 \in \mathbb{X}, \end{cases} \tag{9.7}$$

where $\mathbb{D}_t^\alpha$ is the Caputo fractional derivative of order α, $A \in \Sigma_\omega^\gamma(\mathbb{X})$ with $0 < \omega < \frac{\pi}{2}$, and $f : [0, \infty) \times \mathbb{X} \mapsto \mathbb{X}$ is a jointly continuous function.

Suppose that $A \in \Sigma_\omega^\gamma(\mathbb{X})$ such $-1 < \gamma < 0$ and $0 < \omega < \pi/2$. Define operator families $\{\mathscr{S}_\alpha(t)\}|_{t \in S_{\frac{\pi}{2}-\omega}^0}$, $\{\mathscr{T}_\alpha(t)\}|_{t \in S_{\frac{\pi}{2}-\omega}^0}$ by

$$\mathscr{S}_\alpha(t) := E_\alpha(-zt^\alpha)(A) = \frac{1}{2\pi i} \int_{\Gamma_\theta} E_\alpha(-zt^\alpha)(zI - A)^{-1}dz,$$

$$\mathscr{T}_\alpha(t) := e_\alpha(-zt^\alpha)(A) = \frac{1}{2\pi i} \int_{\Gamma_\theta} e_\alpha(-zt^\alpha)(zI - A)^{-1}dz,$$

where the integral contour $\Gamma_\theta := \{\mathbb{R}_+ e^{i\theta}\} \cup \{\mathbb{R}_+ e^{-i\theta}\}$ is oriented counter-clockwise and $\omega < \theta < \mu < \frac{\pi}{2} - |\arg t|$.

We have

Theorem 9.16 ([110]) *For each fixed $t \in S_{\frac{\pi}{2}-\omega}^0$, $\mathscr{S}_\alpha(t)$ and $\mathscr{T}_\alpha(t)$ are linear and bounded operators on $\mathbb{X}$. Moreover, there exist constants $C_s = C(\alpha, \gamma) > 0$, $C_p = C(\alpha, \gamma) > 0$ such that for all $t > 0$,*

$$\|\mathscr{S}_\alpha(t)\| \leq C_s t^{-\alpha(1+\gamma)}, \quad \|\mathscr{T}_\alpha(t)\| \leq C_p t^{-\alpha(1+\gamma)}. \tag{9.8}$$

9.4.2 Existence Results

Definition 9.17 ([110]) A continuous function $u : (0, T] \mapsto \mathbb{X}$ is called a mild solution to Eq. (9.7) if it satisfies,

$$u(t) = \mathscr{S}_\alpha(t)u_0 + \int_0^t (t-s)^{\alpha-1} \mathscr{T}_\alpha(t-s) f(s, u(s))ds$$

for all $t \in (0, T]$.

Theorem 9.18 ([110]) *Let $A \in \Sigma_\omega^\gamma(\mathbb{X})$ such that $-1 < \gamma < -\frac{1}{2}$ and $0 < \omega < \frac{\pi}{2}$. Suppose that $f : (0, T] \times \mathbb{X} \to \mathbb{X}$ is continuous with respect to t and that there exist constants $M, N > 0$ such that*

$$\|f(t, x) - f(t, y)\| \leq M(1 + \|x\|^{\nu-1} + \|y\|^{\nu-1})\|x - y\|,$$

$$\|f(t, x)\| \leq N(1 + \|x\|^\nu),$$

for all $t \in (0, T]$ and for each $x, y \in \mathbb{X}$, where ν is a constant in $[1, -\frac{\gamma}{1+\gamma})$. Then, for every $u_0 \in \mathbb{X}$, there exists a $T_0 > 0$ such that Eq. (9.7) has a unique mild solution defined on $(0, T_0]$.

Proof Fix $r > 0$ and consider the metric space $(F_r(T, u_0), \rho_T)$ where

$$F_r(T, u_0) = \left\{ u \in C((0, T]; \mathbb{X}) : \; \rho_T(u, \mathscr{S}_\alpha(t)u_0) \leq r \right\},$$

$$\rho_T(u_1, u_2) = \sup_{t \in (0,T]} \|u_1(t) - u_2(t)\|.$$

It can be shown that it is not difficult to see that the metric space $(F_r(T, u_0), \rho_T)$ is complete.

Now for all $u \in F_r(T, u_0)$,

$$\|s^{\alpha(1+\gamma)}u(s)\| \leq s^{\alpha(1+\gamma)}\|u - \mathscr{S}_\alpha(t)u_0\| + s^{\alpha(1+\gamma)}\|\mathscr{S}_\alpha(t)u_0\| \leq L.$$

where $L := T^{\alpha(1+\gamma)}r + C_s\|u_0\|$.

Let $T_0 \in (0, T]$ such that

$$C_p N \frac{T_0^{-\alpha\gamma}}{-\alpha\gamma} + C_p N L^\nu T_0^{-\alpha(\nu(1+\gamma)+\gamma)}\beta(-\gamma\alpha, 1 - \nu\alpha(1 + \gamma)) \leq r, \tag{9.9}$$

$$MC_p \frac{T_0^{-\alpha\gamma}}{-\alpha\gamma} + 2L^{\rho-1}T_0^{-\alpha(\gamma+(1+\gamma)(\nu-1))}\beta(-\alpha\gamma, 1 - \alpha(1 + \gamma)(\nu - 1)) \leq \frac{1}{2}, \tag{9.10}$$

where $\beta(\eta_1, \eta_2)$ with $\eta_i > 0$, $i = 1, 2$ stands for the usual Beta function.

Suppose $u_0 \in \mathbb{X}$ and consider the mapping Γ^α given by

$$(\Gamma^\alpha u)(t) = \mathscr{S}_\alpha(t)u_0 + \int_0^t (t - s)^{\alpha-1}\mathscr{P}_\alpha(t - s)f(s, u(s))ds, \quad u \in F_r(T_0, u_0).$$

From the assumptions upon f, Theorem 9.16, and [110, Theorem 3.2], we deduce that $(\Gamma^\alpha u)(t) \in C((0, T]; \mathbb{X})$ and

$$\|(\Gamma^\alpha u)(t) - \mathscr{S}_\alpha(t)u_0\|$$

$$\leq C_p N \int_0^t (t - s)^{-\alpha\gamma-1}(1 + \|u(s)\|^\nu)ds$$

$$\leq C_p N \frac{T_0^{-\alpha\gamma}}{-\alpha\gamma} + \int_0^t C_p N L^\nu(t - s)^{-\alpha\gamma-1}s^{-\nu\alpha(1+\gamma)}ds$$

$$\leq C_p N \frac{T_0^{-\alpha\gamma}}{-\alpha\gamma} + C_p N L^\nu T_0^{-\alpha(\nu(1+\gamma)+\gamma)}\beta(-\gamma\alpha, 1 - \nu\alpha(1 + \gamma))$$

$$\leq r,$$

by using Eq. (9.9).

In view of the above, one can see that Γ^α maps $F_r(T_0, u_0)$ into itself.

Now for all $u, v \in F_r(T_0, u_0)$, using the assumptions upon f and Theorem 9.16 we deduce that

$$\|(\Gamma^\alpha u)(t) - (\Gamma^\alpha v)(t)\|$$

$$\leq C_p M \int_0^t (t-s)^{-\alpha\gamma-1}(1 + \|u(s)\|^{\rho-1} + \|v(s)\|^{\rho-1})\|u(s) - v(s)\| ds$$

$$\leq C_p M \rho_t(u, v) \int_0^t (t-s)^{-\alpha\gamma-1}(1 + 2L^{\nu-1}s^{-\alpha(\nu-1)(1+\gamma)}) ds$$

$$\leq 2L^{\rho-1} T_0^{-\alpha(\gamma+(1+\gamma)(\nu-1))} \beta(-\alpha\gamma, 1 - \alpha(1+\gamma)(\nu-1)) \rho_{T_0}(u, v)$$

$$+ MC_p \frac{T_0^{-\alpha\gamma}}{-\alpha\gamma} \rho_{T_0}(u, v).$$

Using (9.10), one can easily see that Γ^α is a strict contraction on $F_r(T_0, u_0)$ and so Γ_α has a unique fixed point $u \in F_r(T_0, u_0)$ which, by the Banach Fixed Point Theorem, is the only mild solution to Eq. (9.7) on $(0, T_0]$.

It can be shown that $\mathbb{X}^1 = D(A)$ equipped with the norm defined by $\|x\|_{\mathbb{X}^1} = \|Ax\|$ for all $x \in \mathbb{X}^1$, is a Banach space.

Theorem 9.19 ([110]) *Let $A \in \Theta_\omega^\gamma(\mathbb{X})$ with $-1 < \gamma < -\frac{1}{2}, 0 < \omega < \frac{\pi}{2}$ and $u_0 \in \mathbb{X}^1$. Suppose there exists a continuous function $M_f(\cdot) : \mathbb{R}^+ \to \mathbb{R}^+$ and a constant $N_f > 0$ such that the mapping $f : (0, T] \times \mathbb{X}^1 \to \mathbb{X}^1$ satisfies*

$$\|f(t, x) - f(t, y)\|_{\mathbb{X}^1} \leq M_f(r)\|x - y\|_{\mathbb{X}^1},$$

$$\|f(t, \mathcal{S}_\alpha(t)u_0)\|_{\mathbb{X}^1} \leq N_f(1 + t^{-\alpha(1+\gamma)}\|u_0\|_{\mathbb{X}^1}),$$

for all $0 < t \leq T$ and for all $x, y \in \mathbb{X}^1$ satisfying

$$\sup_{t \in (0,T]} \|x(t) - \mathcal{S}_\alpha(t)u_0\|_{\mathbb{X}^1} \leq r, \quad \sup_{t \in (0,T]} \|y(t) - \mathcal{S}_\alpha(t)u_0\|_{\mathbb{X}^1} \leq r.$$

Then there exists a $T_0 > 0$ such that Eq. (9.7) has a unique mild solution defined on $(0, T_0]$.

Proof Fix $u_0 \in \mathbb{X}^1$ and $r > 0$ and consider

$$F_r''(T, u_0) = \{u \in C((0, T]; \mathbb{X}^1); \sup_{t \in (0,T]} \|u - \mathcal{S}_\alpha(t)u_0\|_{\mathbb{X}^1} \leq r\}.$$

For any $u \in F_r''(T, u_0)$, using the assumptions upon f and Theorem 9.16, we obtain

$$\|(\Gamma^\alpha u)(t) - \mathscr{S}_\alpha(t)u_0\|_{\mathbb{X}^1}$$

$$\leq \int_0^t (t-s)^{\alpha-1}\|\mathscr{P}_\alpha(t-s)\|\|f(s, u(s)) - f(s, \mathscr{S}_\alpha(t)u_0)\|_{\mathbb{X}^1}ds$$

$$+ \int_0^t (t-s)^{\alpha-1}\|\mathscr{P}_\alpha(t-s)\|\|f(s, \mathscr{S}_\alpha(t)u_0))\|_{\mathbb{X}^1}ds$$

$$\leq C_p \int_0^t (t-s)^{-\alpha\gamma-1}(M_f(r)r + N_f + N_f s^{-\alpha(1+\gamma)}\|u_0\|)ds$$

$$\leq C_p(M_f(r)r + N_f)\frac{T^{-\alpha\gamma}}{-\alpha\gamma} + C_p N_f T^{-\alpha(1+2\gamma)}\beta(-\gamma\alpha, 1 - \alpha(1+\gamma))\|u_0\|.$$

In view of the above and ideas from the proof of Theorem 9.18, we obtain the desired result.

9.5   Exercises

1. Show that if A generates an $(\alpha, \alpha)^\beta$-resolvent family T_α^β, then it generates an $(\alpha, 1)^\beta$-resolvent family S_α^β. Further,

$$S_\alpha^\beta(t)x = (g_{1-\alpha} * T_\alpha^\beta)(t)x, \quad t \geq 0, \quad x \in \mathbb{X}.$$

2. Let $A : D(A) \subset \mathbb{X} \mapsto \mathbb{X}$ be a closed linear operator and let $\alpha \in (0, 1]$ and $\beta \geq 0$. Show that if A generates an $(\alpha, 1)^\beta$-resolvent family S_α^β, then the following hold,

   a. $S_\alpha^\beta(t)(D(A)) \subset D(A)$ and

$$AS_\alpha^\beta(t)x = S_\alpha^\beta(t)Ax$$

   for all $x \in D(A)$ and $t \geq 0$.
   b. For all $x \in D(A)$,

$$S_\alpha^\beta(t)x = g_{\alpha\beta+1}(t)x + \int_0^t g_\alpha(t-s)AS_\alpha^\beta(s)xds, \quad t \geq 0.$$

   c. For all $x \in \mathbb{X}$, $(g_\alpha * S_\alpha^\beta)(t)x \in D(A)$,

$$S_\alpha^\beta(t)x = g_{\alpha\beta+1}(t)x + A\int_0^t g_\alpha(t-s)S_\alpha^\beta(s)xds, \quad t \geq 0.$$

   d. $S_\alpha^\beta(0) = g_{\alpha\beta+1}(0)$; $S_\alpha^\beta(0) = I$ if $\beta = 0$ and $S_\alpha^\beta(0) = 0$ if $\beta > 0$.

3. Let $A : D(A) \subset \mathbb{X} \mapsto \mathbb{X}$ be a closed linear operator and let $\alpha \in (0, 1]$ and $\beta \geq 0$. Show that if A generates an $(\alpha, \alpha)^{\beta}$-resolvent family T_{α}^{β}, then the following hold,

 a. $T_{\alpha}^{\beta}(t)(D(A)) \subset D(A)$ and

$$AT_{\alpha}^{\beta}(t)x = T_{\alpha}^{\beta}(t)Ax$$

 for all $x \in D(A)$ and $t > 0$.
 b. For all $x \in D(A)$,

$$T_{\alpha}^{\beta}(t)x = g_{\alpha(\beta+1)}(t)x + \int_{0}^{t} g_{\alpha}(t - s)AT_{\alpha}^{\beta}(s)x\,ds, \quad t \geq 0.$$

 c. For all $x \in \mathbb{X}$, $(g_{\alpha} * T_{\alpha}^{\beta})(t)x \in D(A)$,

$$T_{\alpha}^{\beta}(t)x = g_{\alpha(\beta+1)}(t)x + A \int_{0}^{t} g_{\alpha}(t - s)T_{\alpha}^{\beta}(s)x\,ds, \quad t > 0.$$

 d. If $\beta > 0$, then for every $x \in \overline{D(A)}$,

$$\frac{1}{\Gamma(\alpha(1 + \beta))} \lim_{t \to 0} t^{1-\alpha(1+\beta)} T_{\alpha}^{\beta}(t)x = x$$

 if $\alpha(1+\beta) < 1$; $T_{\alpha}^{\beta}(0)x = x$ if $\alpha(1+\beta) = 1$; and $T_{\alpha}^{\beta}(0)x = 0$ if $\alpha(1+\beta) > 1$.
 e. If $\alpha(1 + \beta) > 1$, then all the above equalities occur for $t \geq 0$. .

4. Suppose $p \in [1, \infty)$, $\alpha \in (0, 1)$, and $\lambda \in [0, \pi)$. Let A_p be the linear operator defined by $A_p = e^{i\lambda}\Delta_p$ where Δ_p is a realization of the Laplace differential operator on $L^p(\mathbb{R}^d)$. Show that A_p is the generator of an $(\alpha, 1)^{\beta}$ on $L^p(\mathbb{R}^d)$ for all $\beta \geq 0$.

9.6 Comments

The material discussed in this chapter is mainly based upon the following two sources: Keyantuo et al. [75] and Wang et al. [110]. One should mention that the semilinear case is not treated in [75]. Consequently, an interesting question consists of using the same tools as in [75] to study the existence of classical (respectively, mild) solutions for the semilinear fractional Cauchy problem

$$\begin{cases} \mathbb{D}_t^{\alpha}u(t) = Au(t) + F(t, u(t)) \\ u(0) = u_0 \in \mathbb{X} \end{cases} \tag{9.11}$$

where $\mathbb{D}_t^\alpha$ is the fractional derivative of order α in the sense of Caputo, $A : D(A) \subset X \mapsto X$ is a closed linear operator on a complex Banach space $\mathbb{X}$, and $f : \mathbb{R}_+ \times \mathbb{X} \mapsto \mathbb{X}$ is a jointly continuous function satisfying some additional conditions.

For the proofs of the existence results Theorem 9.12 and Corollary 9.13, we refer the reader to Keyantuo et al. [75].

The proofs of Theorems 9.18 and 9.19 are taken from Wang et al. [110]. For additional readings upon these topics, we refer to [4, 6, 21–23, 37, 42], etc.

Chapter 10
Second-Order Semilinear Evolution Equations

10.1 Introduction

This chapter is aimed at studying the existence of almost periodic and asymptotically almost periodic solutions to some classes of second-order semilinear evolution equations. In order to establish these existence results, we make extensive use of various tools including the Banach fixed point theorem, the Leray–Schauder alternative, the Sadovsky fixed theorem, etc.

Thermoelastic plate systems play an important role in many applications. For this reason, they have been, in recent years, of a great interest to many researchers. Among other things, the study of the controllability and stability of those thermoelastic plate systems has been considered by many researchers including [14, 18, 24, 44, 65, 79], and [92]. In Sect. 10.2, we study the existence of almost periodic mild solutions to some thermoelastic plate systems with almost periodic forcing terms using mathematical tools such as evolution families and real interpolation spaces.

The main goal of Sect. 10.3 consists of studying the existence of asymptotically almost periodic solutions to some classes of second-order partial functional-differential equations with unbounded delay. The abstract results will, subsequently, be utilized to study the existence of asymptotically almost periodic solutions to some integro-differential equations, which arise in the theory of heat conduction within fading memory materials.

© Springer Nature Switzerland AG 2018
T. Diagana, *Semilinear Evolution Equations and Their Applications*,
https://doi.org/10.1007/978-3-030-00449-1_10

10.2   Almost Periodic Solutions to Some Thermoelastic Plate Systems

10.2.1   Introduction

In this section, we study the existence of almost periodic solutions to thermoelastic plate systems using tools such as evolution families and real interpolation spaces. For that, our main strategy consists of studying its corresponding abstract version. Next, we subsequently use our obtained abstract results to study the existence of almost periodic solutions to these thermoelastic plate systems with almost periodic coefficients.

Let $\Omega \subset \mathbb{R}^N$ ($N \geq 1$) be a bounded subset, which is sufficiently regular and let $a, b : \mathbb{R} \mapsto \mathbb{R}$ be positive functions. The main concern in this section consists of studying the existence (and uniqueness) of almost periodic mild solutions to thermoelastic plate systems given by

$$\begin{cases} u_{tt} + \Delta^2 u + a(t)\Delta\theta = f_1(t, \nabla u, \nabla\theta), \text{ if } t \in \mathbb{R}, \ x \in \Omega \\ \theta_t - b(t)\Delta\theta - a(t)\Delta u_t = f_2(t, \nabla u, \nabla\theta), \text{ if } x \in \Omega, \\ \theta = u = \Delta u = 0, \quad \text{on } \mathbb{R} \times \partial\Omega \end{cases} \tag{10.1}$$

where u, θ are respectively the vertical deflection and the variation of temperature of the plate, the functions f_1, f_2 are continuous and (globally) Lipschitz, and the symbols ∇ and Δ stand respectively for the first and second differential operators given by, $\nabla u = (u_{x_1}, u_{x_2}, \ldots, u_{x_N})$ and

$$\Delta v = \sum_{j=1}^{N} v_{x_j x_j}.$$

Assuming that the coefficients a, b and the forcing terms f_1, f_2 are almost periodic in the first variable (in $t \in \mathbb{R}$) uniformly in the other ones, it will be shown that Eq. (10.1) has a unique almost periodic mild solution.

Recall that a particular case of Eq. (10.1) was investigated by Leiva et al. [80] in the case when not only the coefficients a, b were constant but also there was no gradient terms in the semilinear terms f_1 and f_2. Consequently, the results of this section can be seen as a natural generalization of the results of Leiva et al.

To study the existence of almost periodic solutions to Eq. (10.1), we first study its corresponding abstract semilinear evolution equation and then use the obtained results to establish our existence results. In order to achieve that, let $\mathbb{H} = L^2(\Omega)$ and let A to be the linear operator defined by

$$D(A) = H^2(\Omega) \cap H_0^1(\Omega) \text{ and } A\varphi = -\Delta\varphi \text{ for each } \varphi \in D(A).$$

Setting

$$x := \begin{pmatrix} u \\ u_t \\ \theta \end{pmatrix},$$

then Eq. (10.1) can be easily recast in $\mathbb{X} := D(A) \times \mathbb{H} \times \mathbb{H}$ in the following form

$$x'(t) = A(t)x(t) + f(t, x(t)), \ t \in \mathbb{R}, \tag{10.2}$$

where $A(t)$ are the time-dependent linear operators defined by

$$A(t) = \begin{pmatrix} 0 & I_{\mathbb{X}} & 0 \\ -A^2 & 0 & a(t)A \\ 0 & -a(t)A & -b(t)A \end{pmatrix} \tag{10.3}$$

whose constant domains D are given by

$$D = D(A(t)) = D(A^2) \times D(A) \times D(A), \ t \in \mathbb{R}.$$

Moreover, the semilinear term f is defined only on $\mathbb{R} \times \mathbb{X}_\alpha$ for some $\frac{1}{2} < \alpha < 1$ by

$$f(t, u, v, \theta) = \begin{pmatrix} 0 \\ f_1(t, \nabla u, \nabla \theta) \\ f_2(t, \nabla u, \nabla \theta) \end{pmatrix},$$

where $\mathbb{X}_\alpha$ is the real interpolation space between $\mathbb{X}$ and $D(A(t))$ given by $\mathbb{X}_\alpha = \mathbb{H}_{1+\alpha} \times \mathbb{H}_\alpha \times \mathbb{H}_\alpha$, with $\mathbb{H}_\alpha = (L^2(\Omega), D(A))_{\alpha,\infty} = L^2(\Omega)^A_{\alpha,\infty}$, and $\mathbb{H}_{1+\alpha}$ is the domain of the part of A in $\mathbb{H}_\alpha$.

In Sect. 10.2.3, we show that the family of operators $A(t)$ given in Eq. (10.3) satisfies the Aquistapace–Terreni condition. The fact that each operator $A(t)$ is sectorial was shown in [80]; however, for the sake of clarity and completeness, a complete proof will be given, as we have to determine the precise constants in order to comply with assumption (H.820) from Chap. 8 of this book. Finally, by applying the abstract result developed in Sect. 8.2 of Chap. 8, we prove that the thermoelastic plate system Eq. (10.1) has a unique almost periodic solution

$$\begin{pmatrix} u \\ \theta \end{pmatrix}$$

in $\mathbb{H}_{1+\alpha} \times \mathbb{H}_\alpha$.

10.2.2 Assumptions on the Coefficients of the Thermoelastic System

Let $a, b : \mathbb{R} \mapsto \mathbb{R}$ be positive functions and let $\Omega \subset \mathbb{R}^N$ ($N \geq 1$) be a bounded subset whose boundary $\partial\Omega$ is sufficiently regular. The main objective here consists of studying Eq. (10.1) in the case when the positive real-valued functions a, b are undervalued by a_0, b_0, and $a, b \in C_b^{0,\mu}(\mathbb{R}) \cap AP(\mathbb{R})$, where u, θ are the vertical deflection and the temperature of the plate. Further, it will be assumed that

$$\max_{t \in \mathbb{R}} b^2(t) < 3(a_0^2 + 1). \tag{10.4}$$

In addition to the above assumptions, we suppose that the functions f_1, f_2 : $\mathbb{R} \times H_0^1(\Omega) \times H_0^1(\Omega) \to L^2(\Omega)$ are defined by

$$f_i(t, u, \theta)(x) = f_i(t, \nabla u(x), \nabla \theta(x)) = \frac{K d_i(t)}{1 + |\nabla u(x)| + |\nabla \theta(x)|}$$

for $x \in \Omega$, $t \in \mathbb{R}$, $i = 1, 2$, where $d_i : \mathbb{R} \mapsto \mathbb{R}$ are almost periodic functions.

It is hard to see that the functions f_i ($i = 1, 2$) are jointly continuous. Further, f_i ($i = 1, 2$) are globally Lipschitz, that is, there exists $L > 0$ such that

$$\left\| f_i(t, u, \theta) - f_i(t, v, \eta) \right\|_{L^2(\Omega)} \leq L \left(\|u - v\|_{H_0^1(\Omega)}^2 + \|\theta - \eta\|_{H_0^1(\Omega)}^2 \right)^{\frac{1}{2}}$$

for all $t \in \mathbb{R}, u, v, \eta$ and $\theta \in H_0^1(\Omega)$.

10.2.3 Existence of Almost Periodic Solutions

In order to apply the results of Chap. 8 to this setting, we need to check that some assumptions hold.

Theorem 10.1 ([20, Baroun, Boulite, Diagana, and Maniar]) *Under previous assumptions, the thermoelastic plate system Eq. (10.1) has a unique almost periodic solution*

$$\begin{pmatrix} u \\ \theta \end{pmatrix}$$

in $\mathbb{H}_{1+\alpha} \times \mathbb{H}_\alpha$, whenever L is small enough.

Proof In order to show that $A(t)$ satisfies the Acquistapace–Terreni conditions, we will proceed in two main steps.

Step 1—Let $0 < \lambda_1 < \lambda_2 < \cdots < \lambda_n \to \infty$ be the sequence eigenvalues of A with each eigenvalue being of a finite multiplicity γ_n equal to the dimension of the corresponding eigenspace and $\{\phi_{n,k}\}$ is a complete orthonormal set of eigenvectors for A. For all $x \in D(A)$ we have

$$Ax = \sum_{n=1}^{\infty} \lambda_n \sum_{k=1}^{\gamma_n} \langle x, \phi_{n,k} \rangle \phi_{n,k} := \sum_{n=1}^{\infty} \lambda_n E_n x,$$

with $\langle \cdot, \cdot \rangle$ being the inner product in $\mathbb{H}$.

Obviously, E_n is a complete family of orthogonal projections in $\mathbb{H}$ and so each $x \in \mathbb{H}$ can be written as

$$x = \sum_{n=1}^{\infty} \sum_{k=1}^{\gamma_n} \langle x, \phi_{n,k} \rangle \phi_{n,k} = \sum_{n=1}^{\infty} E_n x.$$

Consequently, for $\begin{pmatrix} w \\ v \\ \theta \end{pmatrix} \in D(A(t))$, the linear operators $A(t)$ can be rewritten as follows,

$$A(t)z = \begin{pmatrix} 0 & I & 0 \\ -A^2 & 0 & a(t)A \\ 0 & -a(t)A & -b(t)A \end{pmatrix} \begin{pmatrix} w \\ v \\ \theta \end{pmatrix}$$

$$= \begin{pmatrix} v \\ -A^2 w + a(t)A\theta \\ -a(t)Av - b(t)A\theta \end{pmatrix} = \begin{pmatrix} \displaystyle\sum_{n=1}^{\infty} E_n v \\ \displaystyle -\sum_{n=1}^{\infty} \lambda_n^2 E_n w + a(t) \sum_{n=1}^{\infty} \lambda_n E_n \theta \\ \displaystyle -a(t) \sum_{n=1}^{\infty} \lambda_n E_n v - b(t) \sum_{n=1}^{\infty} \lambda_n E_n \theta \end{pmatrix}$$

$$= \sum_{n=1}^{\infty} \begin{pmatrix} 0 & 1 & 0 \\ -\lambda_n^2 & 0 & a(t)\lambda_n \\ 0 & -a(t)\lambda_n & -b(t)\lambda_n \end{pmatrix} \begin{pmatrix} E_n & 0 & 0 \\ 0 & E_n & 0 \\ 0 & 0 & E_n \end{pmatrix} \begin{pmatrix} w \\ v \\ \theta \end{pmatrix}$$

$$= \sum_{n=1}^{\infty} A_n(t) P_n z,$$

where

$$P_n := \begin{pmatrix} E_n & 0 & 0 \\ 0 & E_n & 0 \\ 0 & 0 & E_n \end{pmatrix}, \quad n \geq 1,$$

and

$$A_n(t) := \begin{pmatrix} 0 & 1 & 0 \\ -\lambda_n^2 & 0 & a(t)\lambda_n \\ 0 & -a(t)\lambda_n & -b(t)\lambda_n \end{pmatrix}, \ n \geq 1. \tag{10.5}$$

Obviously, the characteristic equation associated with $A_n(t)$ is given by

$$\lambda^3 + b(t)\lambda_n\lambda^2 + (1 + a(t)^2)\lambda_n^2\lambda + b(t)\lambda_n^3 = 0. \tag{10.6}$$

Rescaling as follows, $\lambda/\lambda_n = -\rho$, Eq. (10.6) can be recast as follows,

$$\rho^3 - b(t)\rho^2 + (1 + a(t)^2)\rho - b(t) = 0. \tag{10.7}$$

Using the well-known Routh–Hurwitz theorem we deduce that the real part of the roots $\rho_1(t)$, $\rho_2(t)$, $\rho_3(t)$ of Eq. (10.7) are positive. By a simple calculation one can also verify that Eq. (10.4) does ensure that the roots ρ_1, ρ_2, and ρ_3 are simple and are uniformly separated. In particular, one root is real and the other ones are complex with imaginary part sufficiently far from 0. Consequently, the eigenvalues of $A_n(t)$ are simple and are given by $\sigma_i(t) = -\lambda_n\rho_i(t), i = 1, 2, 3$. Therefore, the matrices $A_n(t)$ are diagonalizable and can be written as follows,

$$A_n(t) = K_n(t)^{-1}J_n(t)K_n(t), n \geq 1,$$

with

$$K_n(t) = \begin{pmatrix} 1 & 1 & 1 \\ \lambda_n\rho_1(t) & \lambda_n\rho_2(t) & \lambda_n\rho_3(t) \\ \dfrac{a(t)\rho_1(t)}{\rho_1(t) - b(t)}\lambda_n & \dfrac{a(t)\rho_2(t)}{\rho_2(t) - b(t)}\lambda_n & \dfrac{a(t)\rho_3(t)}{\rho_3(t) - b(t)}\lambda_n \end{pmatrix},$$

$$J_n(t) = \begin{pmatrix} -\lambda_n\rho_1(t) & 0 & 0 \\ 0 & -\lambda_n\rho_2(t) & 0 \\ 0 & 0 & -\lambda_n\rho_3(t) \end{pmatrix}$$

and

$$K_n(t)^{-1} = \frac{1}{a(a(t), b(t))\lambda_n} \begin{pmatrix} a_{11}(t) & -a_{12}(t) & a_{13}(t) \\ -a_{21}(t) & a_{22}(t) & -a_{23}(t) \\ a_{31}(t) & -a_{32}(t) & a_{33}(t) \end{pmatrix},$$

where

$$a_{11}(t) = \frac{a(t)\rho_3(t)\rho_2(t)(\rho_2(t) - \rho_3(t))}{(\rho_3(t) - b(t))(\rho_2(t) - b(t))}, \quad a_{12}(t) = \frac{a(t)\rho_3(t)\rho_1(t)(\rho_1(t) - \rho_3(t))}{(\rho_3(t) - b(t))(\rho_1(t) - b(t))},$$

$$a_{13}(t) = \frac{a(t)\rho_2(t)\rho_1(t)(\rho_1(t) - \rho_2(t))}{(\rho_2(t) - b(t))(\rho_1(t) - b(t))}, \quad a_{21}(t) = \frac{a(t)b(t)(\rho_2(t) - \rho_3(t))}{(\rho_3(t) - b(t))(\rho_2(t) - b(t))},$$

$$a_{22}(t) = \frac{a(t)b(t)(\rho_1(t) - \rho_3(t))}{(\rho_3(t) - b(t))(\rho_1(t) - b(t))}, \quad a_{23}(t) = \frac{a(t)b(t)(\rho_1(t) - \rho_2(t))}{(\rho_2(t) - b(t))(\rho_1(t) - b(t))},$$

$$a_{31} = (\rho_3(t) - \rho_2(t)), \qquad a_{32} = (\rho_3(t) - \rho_1(t)), \qquad a_{33} = (\rho_2(t) - \rho_1(t)),$$

$$a(a(t), b(t)) = \frac{a(t)\rho_3(t)\rho_2(t)}{(\rho_3(t) - b(t))} + \frac{a(t)\rho_1(t)\rho_3(t)}{(\rho_1(t) - b(t))} + \frac{a(t)\rho_2(t)\rho_1(t)}{(\rho_2(t) - b(t))}$$

$$- \frac{a(t)\rho_1(t)\rho_2(t)}{(\rho_1(t) - b(t))} - \frac{a(t)\rho_3(t)\rho_1(t)}{(\rho_3(t) - b(t))} - \frac{a(t)\rho_2(t)\rho_3(t)}{(\rho_2(t) - b(t))}.$$

From the fact that $b(\cdot)$ is not a solution to Eq. (10.7), it can be shown that the matrix operators $K_n(t)$ and $K_n^{-1}(t)$ are well defined and $K_n(t)P_n(t) : Z := \mathbb{H} \times \mathbb{H} \times \mathbb{H} \mapsto \mathbb{X}$, $K_n^{-1}(t)P_n(t) : \mathbb{X} \mapsto Z$.

We claim that the roots $\rho_i(t), i = 1, 2, 3$, of Eq. (10.7) are bounded. Indeed, setting $l(t) = \rho(t) - \frac{b(t)}{3}$, then Eq. (10.7) becomes

$$l(t)^3 + p(t)l(t) + q(t) = 0,$$

where $p(t) := (1 + a(t)^2) - \frac{b(t)^2}{3}$, $q(t) := \frac{2}{27}b(t)^3 - (2 - a(t)^2)\frac{b(t)}{3}$. Since q is bounded and

$$|q(t)| = |l(t)||l(t)^2 + p(t)| \geq |l(t)||l(t)|^2 - |p(t)|,$$

then l is also bounded. Thus the boundedness of b yields the above claim.

Define the sector S_θ as

$$S_\theta = \{\lambda \in \mathbb{C} : |\arg(\lambda)| \leq \theta, \ \lambda \neq 0\},$$

where

$$0 \leq \sup_{t \in \mathbb{R}} |\arg(\rho_i(t))| < \frac{\pi}{2}, \ i = 1, 2, 3$$

and

$$\frac{\pi}{2} < \theta < \pi - \max_{i=1,2,3} \sup_{t \in \mathbb{R}} \{|\arg(\rho_i(t))|\}.$$

For $\lambda \in S_\theta$ and $z \in \mathbb{X}$, one has

$$R(\lambda, A(t))z = \sum_{n=1}^{\infty} (\lambda - A_n(t))^{-1} P_n z$$

$$= \sum_{n=1}^{\infty} K_n(t)(\lambda - J_n(t)P_n)^{-1} K_n^{-1}(t) P_n z.$$

Hence,

$$\|R(\lambda, A(t))z\|^2 \leq \sum_{n=1}^{\infty} \|K_n(t)P_n(\lambda - J_n(t)P_n)^{-1} K_n^{-1}(t) P_n\|_{B(\mathbb{X})}^2 \|P_n z\|^2$$

$$\leq \sum_{n=1}^{\infty} \|K_n(t)P_n\|_{B(\mathbb{Z},\mathbb{X})}^2 \|(\lambda - J_n(t)P_n)^{-1}\|_{B(\mathbb{Z})}^2$$

$$\cdot \|K_n^{-1}(t)P_n\|_{B(\mathbb{X},\mathbb{Z})}^2 \|P_n z\|^2.$$

Using Eq. (10.7) and the fact that $b(t) > b_0$ par assumption, it follows that

$$|\rho(t) - b(t)| \geq \frac{a(t)^2 |\rho(t)|}{1 + |\rho(t)|^2}, \qquad \inf_{t \in \mathbb{R}} |\rho(t)| > 0. \tag{10.8}$$

Consequently, from the assumption $a(t) > a_0$ it follows that

$$\inf_{t \in \mathbb{R}} |\rho(t) - b(t)| > 0. \tag{10.9}$$

Moreover, for $z := \begin{pmatrix} z_1 \\ z_2 \\ z_3 \end{pmatrix} \in Z$, we have

$$\|K_n(t)P_n z\|^2 = \lambda_n^2 \|E_n z_1 + E_n z_2 + E_n z_3\|^2$$

$$+ \lambda_n^2 \|\rho_1(t)E_n z_1 + \rho_2(t)E_n z_2 + \rho_3(t)E_n z_3\|^2$$

$$+ \lambda_n^2 \left\| \frac{a(t)\rho_1(t)}{\rho_1(t) - b(t)} E_n z_1 + \frac{a(t)\rho_2(t)}{\rho_2(t) - b(t)} E_n z_2 + \frac{a(t)\rho_3(t)}{\rho_3(t) - b(t)} E_n z_3 \right\|^2.$$

Therefore, there is $C_1 > 0$ such that

$$\|K_n(t)P_n z\|_H \leq C_1 \lambda_n \|z\|_Z \quad \text{for all} \quad n \geq 1 \text{ and } t \in \mathbb{R}.$$

Arguing as above, for $z := \begin{pmatrix} z_1 \\ z_2 \\ z_3 \end{pmatrix} \in \mathbb{X}$, one can show

$$\|K_n^{-1}(t) P_n z\| \le \frac{C_2}{\lambda_n} \|z\| \quad \text{for all } n \ge 1 \text{ and } t \in \mathbb{R}.$$

Now, for $z \in Z$, we have

$$\|(\lambda - J_n P_n)^{-1} z\|_Z^2 = \left\| \begin{pmatrix} \frac{1}{\lambda + \lambda_n \rho_1(t)} & 0 & 0 \\ 0 & \frac{1}{\lambda + \lambda_n \rho_2(t)} & 0 \\ 0 & 0 & \frac{1}{\lambda + \lambda_n \rho_3(t)} \end{pmatrix} \begin{pmatrix} z_1 \\ z_2 \\ z_3 \end{pmatrix} \right\|_Z^2$$

$$\le \frac{1}{(\lambda + \lambda_n \rho_1(t))^2} \|z_1\|^2 + \frac{1}{(\lambda + \lambda_n \rho_2(t))^2} \|z_2\|^2$$

$$+ \frac{1}{(\lambda + \lambda_n \rho_3(t))^2} \|z_3\|^2.$$

Let $\lambda_0 > 0$. Obviously, the function defined by

$$\eta(\lambda) := \frac{1 + |\lambda|}{|\lambda + \lambda_n \rho_i(t)|}$$

is continuous and bounded on the closed set $\Sigma := \{\lambda \in \mathbb{C} / |\lambda| \le \lambda_0, \ |\arg \lambda| \le \theta\}$.

On the other hand, it is clear that η is bounded for $|\lambda| > \lambda_0$. Thus η is bounded on S_θ. If we take

$$N = \sup \left\{ \frac{1 + |\lambda|}{|\lambda + \lambda_n \rho_i(t)|} \ : \ \lambda \in S_\theta, n \ge 1; \ i = 1, 2, 3, \ t \in \mathbb{R} \right\}.$$

Therefore,

$$\|(\lambda - J_n P_n)^{-1} z\|_Z \le \frac{N}{1 + |\lambda|} \|z\|_Z, \quad \lambda \in S_\theta.$$

Consequently,

$$\|R(\lambda, A(t))\| \le \frac{K}{1 + |\lambda|}$$

for all $\lambda \in S_\theta$ and $t \in \mathbb{R}$.

The operators $A(t)$ are invertible and their inverses are given by

$$A(t)^{-1} = \begin{pmatrix} -a(t)^2 b(t)^{-1} A^{-1} -A^{-2} & -a(t)b(t)^{-1}A^{-2} \\ I & 0 & 0 \\ -a(t)b(t)^{-1} & 0 & -b(t)^{-1}A^{-1} \end{pmatrix}, \quad t \in \mathbb{R}.$$

Hence, for $t, s, r \in \mathbb{R}$, one has

$$(A(t) - A(s))A(r)^{-1}$$

$$= \begin{pmatrix} 0 & 0 & 0 \\ -a(r)b(r)^{-1}(a(t) - a(s))A & 0 & -b(r)^{-1}(a(t) - a(s)) \\ -(a(t) - a(s))A + a(r)b(r)^{-1}(b(t) - b(s))A & 0 & -b(r)^{-1}(b(t) - b(s)) \end{pmatrix},$$

and hence

$$\begin{aligned}
\|(A(t) - A(s))A(r)^{-1}z\| &\leq \sqrt{3}(\|a(r)b(r)^{-1}(a(t) - a(s))Az_1\| \\
&\quad + \|b(r)^{-1}(a(t) - a(s))z_3\| + \|(a(t) - a(s))Az_1\| \\
&\quad + \|a(r)b(r)^{-1}(b(t) - b(s))Az_1\| \\
&\quad + \|b(r)^{-1}(b(t) - b(s))z_3\|) \\
&\leq \sqrt{3}(|a(r)b(r)^{-1}|\,|t - s|^{\mu}\|Az_1\| + |b(r)^{-1}|\,|t - s|^{\mu}\|z_3\| \\
&\quad + |t - s|^{\mu}\|Az_1\| + |a(r)b(r)^{-1}|\,|t - s|^{\mu}\|Az_1\| \\
&\quad + |b(r)^{-1}|\,|t - s|^{\mu}\|z_3\|) \\
&\leq (2\sqrt{3}|a(r)b(r)^{-1}| + 1)|t - s|^{\mu}\|Az_1\| \\
&\quad + 2\sqrt{3}|a(r)b(r)^{-1}|\,|t - s|^{\mu}\|z_3\|.
\end{aligned}$$

Consequently,

$$\|(A(t) - A(s))A(r)^{-1}z\| \leq C|t - s|^{\mu}\|z\|.$$

Step 2—For every $t \in \mathbb{R}$, $A(t)$ generates an analytic semigroup $(e^{\tau A(t)})_{\tau \geq 0}$ on $\mathbb{X}$. Using similar computations as above, one can show that

$$\sup_{t,s \in \mathbb{R}} \|A(t)A(s)^{-1}\| < \infty$$

and for every $t, s \in \mathbb{R}$ and $0 < \mu \leq 1$,

$$\|A(t)A(s)^{-1} - Id\| \leq L'k|t - s|^{\mu}$$

with constant $L' \geq 0$ and k is the Lipschitz constant of the functions a and b.

On the other hand, we have

$$e^{\tau A(t)}z = \sum_{n=0}^{\infty} K_n(t)^{-1} P_n e^{\tau J_n} P_n K_n(t) P_n z, \quad z \in \mathbb{X}.$$

Then,

$$\|e^{\tau A(t)}z\| = \sum_{n=0}^{\infty} \|K_n(t)^{-1} P_n\|_{B(\mathbb{X},Z)} \|e^{\tau J_n} P_n\|_{B(Z)} \|K_n(t) P_n\|_{B(Z,\mathbb{X})} \|P_n z\|,$$

with for each $z = \begin{pmatrix} z_1 \\ z_2 \\ z_3 \end{pmatrix} \in Z$

$$
\begin{aligned}
\|e^{\tau J_n} P_n z\|_Z^2 &= \left\| \begin{pmatrix} e^{-\lambda_n \rho_1(t)\tau} E_n & 0 & 0 \\ 0 & e^{-\lambda_n \rho_2(t)\tau} E_n & 0 \\ 0 & 0 & e^{-\lambda_n \rho_3(t)\tau} E_n \end{pmatrix} \begin{pmatrix} z_1 \\ z_2 \\ z_3 \end{pmatrix} \right\|_Z^2 \\
&\leq \|e^{-\lambda_n \rho_1(t)\tau} E_n z_1\|^2 + \|e^{-\lambda_n \rho_2(t)\tau} E_n z_2\|^2 + \|e^{-\lambda_n \rho_3(t)\tau} E_n z_3\|^2 \\
&\leq e^{-2\delta\tau} \|z\|_Z^2,
\end{aligned}
$$

where $\delta = \lambda_1 \inf_{t \in \mathbb{R}} \{Re(\rho_1(t)), Re(\rho_2(t)), Re(\rho_3(t))\}$.
  Therefore

$$\|e^{\tau A(t)}\| \leq C e^{-\delta\tau}, \quad \tau \geq 0. \tag{10.10}$$

Using the continuity of the functions a, b and the spectral identity

$$R(\lambda, A(t)) - R(\lambda, A(s)) = R(\lambda, A(t))(A(t) - A(s))R(\lambda, A(s))$$

it follows that the mapping $J \ni t \mapsto R(\lambda, A(t))$ is strongly continuous for $\lambda \in S_\theta$ where $J \subset \mathbb{R}$ is an arbitrary compact interval. Therefore, $A(t)$ satisfies the assumptions of [104, Corollary 2.3] and thus, the evolution family $U(t, s)$ is exponentially stable. The step 2 is complete.
  To complete the proof, we have to show $(A(\cdot))^{-1} \in AP(\mathbb{R}, B(\mathbb{X}))$ (See assumption (H.823) of [Sect. 8.3, Chap. 8]). Let $\epsilon > 0$, and $\tau = \tau_\epsilon \in \mathscr{P}(\epsilon, a, b)$. We have

$$A(t)^{-1} - A(t + \tau)^{-1} = A(t + \tau)^{-1}(A(t + \tau) - A(t))A(t)^{-1}, \tag{10.11}$$

and,

$$A(t + \tau) - A(t) = \begin{pmatrix} 0 & 0 & 0 \\ 0 & 0 & (a(t + \tau) - a(t))A \\ 0 & -(a(t + \tau) - a(t))A & -(b(t + \tau) - b(t))A \end{pmatrix}.$$

Therefore, for $z := \begin{pmatrix} z_1 \\ z_2 \\ z_3 \end{pmatrix} \in D$, one has

$$\begin{aligned} \|(A(t + \tau) - A(t))z\| &\leq \|(a(t + \tau) - a(t))Az_3\| + \|(a(t + \tau) - a(t))Az_2\| \\ &\quad + \|(b(t + \tau) - b(t))Az_3\| \\ &\leq \varepsilon\|Az_2\| + \varepsilon\|Az_3\| \\ &\leq \varepsilon\|z\|_D, \end{aligned}$$

and using Eq. (10.11) ($\|\cdot\|_D$ being the graph norm with respect to the domain $D = D(A(t))$), we obtain

$$\begin{aligned} \|A(t + \tau)^{-1}y - A(t)^{-1}y\| &\leq \|A(t + \tau)^{-1}(A(t + \tau) - A(t))A(t)^{-1}y\| \\ &\leq \|A(t + \tau)^{-1}\|_{B(\mathbb{X})} \\ &\quad \times \|(A(t + \tau) - A(t))\|_{B(D,\mathbb{X})}\|A(t)^{-1}y\|_D, \quad y \in \mathbb{X}. \end{aligned}$$

Since $\|A(t)^{-1}y\|_D \leq c\|y\|$, then

$$\|A(t + \tau)^{-1}y - A(t)^{-1}y\| \leq c'\varepsilon\|y\|.$$

Consequently, $A(t)^{-1}$ is almost periodic.

Finally, for L sufficiently small, all assumptions of Theorem 8.21 are satisfied and thus the thermoelastic system Eq. (10.1) has a unique almost periodic mild solution

$$\begin{pmatrix} u \\ \theta \end{pmatrix}$$

with values in the interpolation space $\mathbb{H}_{1+\alpha} \times \mathbb{H}_\alpha$.

10.3 Existence Results for Some Second-Order Partial Functional Differential Equations

The main focus in this section consists of studying the existence of asymptotically almost periodic solutions to some classes of second-order partial functional differential equations with unbounded delay. The abstract results will, subsequently, be utilized to studying the existence of asymptotically almost periodic solutions to some integro-differential equations which arise in the theory of heat conduction within fading memory materials.

10.3.1 Introduction

Our main concern in this section consists of studying the existence of asymptotically almost periodic solutions to the class of second-order abstract partial functional differential equations of the form

$$\frac{d}{dt}\left[x'(t) - g(t, x_t)\right] = Ax(t) + f(t, x_t), \qquad t \in I, \tag{10.12}$$

$$x_0 = \varphi \in \mathscr{B}, \tag{10.13}$$

$$x'(0) = \xi \in \mathbb{X}, \tag{10.14}$$

where A is the infinitesimal generator of a strongly continuous cosine family $(C(t))_{t \in \mathbb{R}}$ of bounded linear operators on $\mathbb{X}$, the history $x_t : (-\infty, 0] \to \mathbb{X}$, $x_t(\theta) := x(t + \theta)$, belongs to an abstract phase space $\mathscr{B}$ defined axiomatically, and f, g are some appropriate functions.

Recall that the abstract Cauchy systems of the form, Eqs. (10.12)–(10.14) arise, for instance, in the theory of heat conduction in materials with fading memories, see, e.g., Gurtin–Pipkin [62] and Nunziato [96]. In the classical theory of heat conduction, it is assumed that the internal energy and the heat flux depend linearly upon the temperature u as well as its gradient ∇u. Under these conditions, the classical heat equation describes sufficiently well the evolution of the temperature in different types of materials. However, this description is not satisfactory for materials with fading memories. In the theory developed in [62, 96], the internal energy and the heat flux are described as functionals of u and u_x. Upon some physical conditions, they established that the temperature $u(t, \xi)$ satisfies the integro-differential equation

$$c\frac{\partial^2 u(t, \xi)}{\partial t^2} = \beta(0)\frac{\partial u(t, \xi)}{\partial t} + \int_0^\infty \beta'(s)\frac{\partial u(t - s, \xi)}{\partial t}ds + \alpha(0)\Delta u(t, \xi)$$

$$+ \int_0^\infty \alpha'(s)\Delta u(t - s, \xi)ds, \tag{10.15}$$

where $\beta(\cdot)$ is the energy relaxation function, $\alpha(\cdot)$ is the stress relaxation function and c is the density. Assuming that $\beta(\cdot)$ is smooth enough and that $\nabla u(t, \xi)$ is approximately constant at t, we can rewrite the previous equation in the form

$$\frac{\partial^2 u(t, \xi)}{\partial t^2} = \frac{\partial}{\partial t} \left[\frac{\beta(0)}{c} u(t, \xi) + \frac{1}{c} \int_0^\infty \beta'(s) u(t - s, \xi) ds \right] + d\Delta u(t, \xi).$$

By making the function $\beta(\cdot)$ explicitly dependent on the time t, we can consider the situation in which the material is submitted to an aging process so that the hereditary properties are lost as the time goes to infinity. In this case, the previous equation takes the form

$$\frac{\partial^2 u(t, \xi)}{\partial t^2} = \frac{\partial}{\partial t} \left[\frac{\beta(t, 0)}{c} u(t, \xi) + \frac{1}{c} \int_0^\infty \frac{\partial \beta(t, s)}{\partial s} u(t - s, \xi) ds \right] \quad (10.16)$$

$$+ d\Delta u(t, \xi),$$

which can be transformed into the abstract systems of the form Eqs. (10.12)–(10.14) assuming that the solution $u(\cdot)$ is known on $[0, \infty)$.

10.3.2 Preliminaries and Notations

In the rest of this section, if W is an arbitrary metric space, then the notation $B_r(x, W)$ stands for the closed ball in W, centered at x with radius r. The linear operator $A : D(A) \subset \mathbb{X} \to \mathbb{X}$ considered here will be assumed to be the infinitesimal generator of a strongly continuous cosine family $(C(t))_{t \in \mathbb{R}}$ of bounded linear operators on $\mathbb{X}$ and $(S(t))_{t \in \mathbb{R}}$ denote the associated sine function, which is defined by

$$S(t)x = \int_0^t C(s)x ds, \mathscr{H} x \in \mathbb{X}, t \in \mathbb{R}.$$

For further details upon cosine function theory and their applications to the second-order abstract Cauchy problem, we refer the reader to Fattorini [57] and Travis and Webb [106, 107].

Recall that Travis and Webb [106] studied the existence of solutions to the second-order abstract Cauchy problem,

$$x''(t) = Ax(t) + h(t), \quad t \in [0, b], \quad (10.17)$$

$$x(0) = w, \quad x'(0) = z, \quad (10.18)$$

where $h \in L^1([0, b]; \mathbb{X})$.

The corresponding semilinear case was also done by Travis and Webb [107].

Recall that a mild solution for the system Eqs. (10.17)–(10.18) is any function x that is given by

$$x(t) = C(t)w + S(t)z + \int_0^t S(t-s)h(s)ds, \quad t \in [0, b]. \qquad (10.19)$$

In this section the definition of the phase space $\mathscr{B}$ will be as in [69]. Namely, $\mathscr{B}$ will be a vector space of functions mapping $(-\infty, 0]$ into $\mathbb{X}$ endowed with a semi-norm $\| \cdot \|_{\mathscr{B}}$. Moreover, we will assume that the following axioms hold,

(A) If $x : (-\infty, \sigma + b] \to \mathbb{X}$, $b > 0$, is such that $x_\sigma \in \mathscr{B}$ and $x|_{[\sigma, \sigma+b]} \in C([\sigma, \sigma + b]; \mathbb{X})$, then for every $t \in [\sigma, \sigma + b)$ the following conditions hold:

   (i) x_t is in $\mathscr{B}$,
   (ii) $\| x(t) \| \le H \| x_t \|_{\mathscr{B}}$,
   (iii) $\| x_t \|_{\mathscr{B}} \le K(t - \sigma) \sup\{\| x(s) \| : \sigma \le s \le t\} + M(t - \sigma) \| x_\sigma \|_{\mathscr{B}}$,

   where $H > 0$ is a constant, $K, M : [0, \infty) \to [1, \infty)$ with K being continuous, M is locally bounded, and H, K, M are independent of $x(\cdot)$.

(A1) For the function $x(\cdot)$ in (A), the function $t \to x_t$ is continuous from $[\sigma, \sigma + a)$ into $\mathscr{B}$.

(B) The space $\mathscr{B}$ is complete.

(C2) If $(\varphi^n)_{n \in \mathbb{N}}$ is a uniformly bounded sequence in $C((-\infty, 0]; \mathbb{X})$ formed by functions with compact support and $\varphi^n \to \varphi$ in the compact-open topology, then $\varphi \in \mathscr{B}$ and

$$\|\varphi^n - \varphi\|_{\mathscr{B}} \to 0$$

as $n \to \infty$.

Example 10.2 (The Phase Space $\mathbf{C_r} \times \mathbf{L^p}(\rho; \mathbb{X}))$ Let $r \ge 0$, $1 \le p < \infty$ and let $\rho : (-\infty, -r] \to \mathbb{R}$ be a nonnegative measurable function which satisfies the conditions (g-5), (g-6) in the terminology of [69]. Briefly, this means that ρ is a locally integrable function and that there exists a nonnegative locally bounded function γ on $(-\infty, 0]$ such that

$$\rho(\xi + \theta) \le \gamma(\xi)\rho(\theta),$$

for all $\xi \le 0$ and $\theta \in (-\infty, -r) \setminus N_\xi$, where $N_\xi \subseteq (-\infty, -r)$ is a set whose Lebesgue measure is zero.

The space $C_r \times L^p(\rho; \mathbb{X})$ consists of all classes of functions $\varphi : (-\infty, 0] \to \mathbb{X}$ such that φ is continuous on $[-r, 0]$, Lebesgue-measurable, and $\rho\|\varphi\|^p$ is Lebesgue integrable on $(-\infty, -r)$. The semi-norm on $C_r \times L^p(\rho; \mathbb{X})$ is defined by

$$\|\varphi\|_{\mathscr{B}} := \sup\left\{\|\varphi(\theta)\| : -r \le \theta \le 0\right\} + \left(\int_{-\infty}^{-r} \rho(\theta)\|\varphi(\theta)\|^p d\theta\right)^{1/p}.$$

The space $\mathscr{B} = C_r \times L^p(\rho; \mathbb{X})$ satisfies axioms **(A)-(A1)-(B)**. Moreover, when $r = 0$ and $p = 2$, we can take

$$H = 1, \quad M(t) = \gamma(-t)^{1/2}, \quad \text{and} \quad K(t) = 1 + \left(\int_{-t}^{0} \rho(\theta)\, d\theta \right)^{1/2}$$

for $t \geq 0$, see [69, Theorem 1.3.8] for details.

Some of our existence results require some additional assumptions upon the phase space $\mathscr{B}$.

Definition 10.3 Let $\mathscr{B}_0 = \{\psi \in \mathscr{B} : \psi(0) = 0\}$. The phase space $\mathscr{B}$ is called a fading memory space if $\| S(t)\psi \|_{\mathscr{B}} \to 0$ as $t \to \infty$ for every $\psi \in \mathscr{B}_0$. We say that $\mathscr{B}$ is a uniform fading memory space if $\|S(t)\|_{\mathscr{L}(\mathscr{B}_0)} \to 0$ as $t \to \infty$.

For further details upon phase spaces, we refer the reader to for instance [69].

Let $I \subset \mathbb{R}$ be an interval. Recall that the spaces $BC(I; \mathscr{V}) = C_b(I; \mathscr{V})$ and $C_0([0, \infty); \mathscr{V})$ are defined respectively by

$$BC(I, \mathscr{V}) = C_b(I; \mathscr{V})$$
$$= \left\{ x : I \to \mathscr{V}, \ x \text{ is continuous and } \|x\| = \sup_{t \in I} \| x(t) \| < \infty \right\},$$
$$C_0([0, \infty); \mathscr{V}) = \left\{ x \in C_b([0, \infty); \mathscr{V}) : \lim_{t \to \infty} \|x(t)\| = 0 \right\},$$

and both spaces are endowed with their corresponding sup-norms.

10.3.3 Existence of Local and Global Mild Solutions

We establish the existence of mild solutions to Eqs. (10.12)–(10.14) in the particular cases when $I = [0, a]$ and $I = [0, \infty)$. Suppose, $I = [0, a]$ or $I = [0, \infty)$ and let $N, \tilde{N}$ be positive constants such that

$$\|C(t)\| \leq N$$

and

$$\|S(t)\| \leq \tilde{N}$$

for every $t \in I$.

Our existence results require the following general assumption,

(H₁)   The functions $f, g : I \times \mathscr{B} \to \mathbb{X}$ satisfy the following conditions:

  (i) The functions $f(t, \cdot), g(t, \cdot) : \mathscr{B} \to \mathbb{X}$ are continuous a.e. $t \in I$.

(ii) For each $\psi \in \mathcal{B}$, the functions $f(\cdot, \psi), g(\cdot, \psi) : I \to \mathbb{X}$ are strongly measurable.

(iii) There exist integrable functions $m_f, m_g : I \to [0, \infty)$ and continuous nondecreasing functions $W_f, W_g : [0, \infty) \to (0, \infty)$ such that

$$\| f(t, \psi) \| \leq m_f(t) W_f(\| \psi \|_{\mathcal{B}}), \quad (t, \psi) \in I \times \mathcal{B},$$

$$\| g(t, \psi) \| \leq m_g(t) W_g(\| \psi \|_{\mathcal{B}}), \quad (t, \psi) \in I \times \mathcal{B}.$$

Motivated by the concept of mild solution given in Eq. (10.19), we adopt the following concept of mild solution for Eqs. (10.12)–(10.14).

Definition 10.4 A function $x : (-\infty, 0] \cup I \to \mathbb{X}$ is called a mild solution of the abstract Cauchy problem Eqs. (10.12)–(10.14) on I, if $x_0 = \varphi$ and

$$x(t) = C(t)\varphi(0) + S(t)[\xi - g(0, \varphi)] + \int_0^t C(t - s)g(s, x_s)ds$$

$$+ \int_0^t S(t - s)f(s, x_s)ds, \quad t \in I.$$

In the rest of this section, we set $W := \max\{W_f, W_g\}$.

10.3.4 Existence of Solutions in Bounded Intervals

Recall that the existence of mild solutions to Eqs. (10.12)–(10.14) in the case when $I = [0, a]$ can be obtained through the results in [67]. However, for the sake of clarity and completeness, we provide the reader with the proof of the next theorem, as some of the ideas in this proof are also needed in the sequel.

Theorem 10.5 *Suppose that assumption* ($\mathbf{H_1}$) *holds and that for every* $0 < t \leq a$ *and* $r > 0$, *the sets* $U(t, r) = \{S(t)f(s, \psi) : s \in [0, t], \| \psi \|_{\mathcal{B}} \leq r\}$ *and* $g(I \times B_r(0, \mathcal{B}))$ *are relatively compact in* $\mathbb{X}$. *If*

$$K_a \int_0^a (Nm_g(s) + \tilde{N}m_f(s)) \, ds < \int_c^\infty \frac{ds}{W(s)}, \tag{10.20}$$

where

$$c = (K_a NH + M_a) \| \varphi \|_{\mathcal{B}} + K_a \tilde{N}(\|\xi\| + \|g(0, \varphi)\|),$$

$$K_a = \sup_{s \in [0, a]} K(s),$$

and

$$M_a = \sup_{s \in [0,a]} M(s),$$

then the system Eqs. (10.12)–(10.14) has a mild solution.

Proof Let the vector space

$$\mathscr{B}C = \left\{ x : (-\infty, a] \to \mathbb{X} : x|_{(-\infty,0]} \in \mathscr{B}, \ x|_{[0,a]} \in C([0,a]; \mathbb{X}) \right\}$$

be endowed with the norm defined by

$$\|x\|_{\mathscr{B}C} = \|x|_{(-\infty,0]}\|_{\mathscr{B}} + \|x|_{[0,a]}\|_a$$

for all $x \in \mathscr{B}C$.

On this space, we define the map $\Gamma : \mathscr{B}C \to \mathscr{B}C$ by $(\Gamma x)_0 = \varphi$ and

$$\Gamma x(t) = C(t)\varphi(0) + S(t)[\xi - g(0,\varphi)] + \int_0^t C(t-s)g(s,x_s)ds$$

$$+ \int_0^t S(t-s)f(s,x_s)ds, \quad t \in I.$$

It is then easy to see that Γx is well defined and that $\Gamma x \in \mathscr{B}C$. Moreover, by using the phase space axioms and the Lebesgue Dominated Convergence Theorem, one can prove that Γ is a continuous function from $\mathscr{B}C$ into $\mathscr{B}C$.

In order to apply Theorem 1.81, we establish an *a priori* estimate for the solution of the integral equation $x = \lambda \Gamma x$, $\lambda \in (0,1)$. Let $x^\lambda \in \mathscr{B}C$ be a solution of $x = \lambda \Gamma x$, $\lambda \in (0,1)$. For $t \in I$, we get

$$\| x^\lambda(t) \| \leq NH \| \varphi \|_{\mathscr{B}} + \widetilde{N}(\| \xi \| + \| g(0,\varphi) \|) + \int_0^t (Nm_g(s)$$

$$+ \widetilde{N}m_f(s))W(\| x_s^\lambda \|_{\mathscr{B}})ds$$

which yields

$$\| x_t^\lambda \|_{\mathscr{B}} \leq (K_a NH + M_a) \| \varphi \|_{\mathscr{B}} + K_a \widetilde{N}(\| \xi \| + \| g(0,\varphi) \|)$$

$$+ K_a \int_0^t (Nm_g(s) + \widetilde{N}m_f(s))W(\| x_s^\lambda \|_{\mathscr{B}})ds.$$

Denoting by $\beta_\lambda(t)$ the right-hand side of the last inequality, we find that

$$\beta_\lambda'(t) \leq K_a(Nm_g(t) + \widetilde{N}m_f(t))W(\beta_\lambda(t)),$$

and hence,

$$\int_{\beta_\lambda(0)=c}^{\beta_\lambda(t)} \frac{ds}{W(s)} \le K_a \int_0^t (Nm_g(s) + \tilde{N}m_f(s))\, ds < \int_c^\infty \frac{ds}{W(s)},$$

which enables to conclude that the set of functions $\{\beta_\lambda : \lambda \in (0, 1)\}$ is bounded. As a consequence of the previous fact, $\{x^\lambda : \lambda \in (0, 1)\}$ is bounded in $C(I, X)$ as

$$\| x^\lambda(t) \| \le H \| x_t^\lambda \| \le \beta_\lambda(t)$$

for every $t \in I$.

On the other hand, from [68, Lemma 3.1] we deduce that Γ is completely continuous on $\mathscr{B}C$. The existence of a mild solution for Eqs. (10.12)–(10.14) is now a consequence of Theorem 1.81.

In many situations of practical interest, the sine function $S(t)$ is compact. This is the motivation for the next result.

Corollary 10.6 *Suppose that assumption* (**H$_1$**) *holds and that $S(t)$ is compact for all $t \ge 0$ and the set $g(I \times B_r(0, \mathscr{B}))$ is relatively compact in $\mathbb{X}$ for every $r > 0$. If Eq. (10.20) holds, then the system Eqs. (10.12)–(10.14) has a mild solution.*

Remark 10.7 Recall that except when the space $\mathbb{X}$ is a finite dimensional space, the cosine function is not compact, and that for this reason, the compactness assumption on the function g cannot be removed. For more on this and related issues, we refer the reader to for instance the work of Travis and Webb [106, pp. 557].

Using similar ideas as in the proof of Theorem 10.5, we can prove the following local existence result.

Theorem 10.8 *Suppose that assumption* (**H$_1$**) *holds and that for every $0 < t \le a$ and $r > 0$, the sets*

$$U(t, r) = \left\{ S(t)f(s, \psi) : s \in [0, t], \|\psi\|_{\mathscr{B}} \le r \right\}$$

and

$$g(I \times B_r(0, \mathscr{B}))$$

are relatively compact in $\mathbb{X}$.

Then there exists a mild solution to Eqs. (10.12)–(10.14) on $[0, b]$ for some $0 < b \le a$.

10.3.5   Existence of Global Solutions

In this subsection, our discussions will be upon the existence of mild solutions defined on the interval $I = [0, \infty)$. For that, we suppose that M, K are positive constants such that $M(t) \leq M$ and $K(t) \leq K$ for every $t \geq 0$ and that the functions m_f, m_g are locally integrable.

We need the following notations

$$W = \max\{W_f, W_g\}, \quad m = \max\{m_f, m_g\}, \quad \gamma(s) = N m_g(s) + \widetilde{N} m_f(s).$$

Remark 10.9 Recall that if $\mathscr{B}$ is a fading memory space, then the functions $M(\cdot)$, $K(\cdot)$ are bounded on $[0, \infty)$. For further details on this and related issues, we refer the reader to [69, Proposition 7.1.5].

Let $h : [0, \infty) \to (0, \infty)$ be a continuous nondecreasing function with $h(0) = 1$ and such that $h(t) \to \infty$ as $t \to \infty$.

Let $C_{0,h}(\mathbb{X})$ denote the space defined by

$$C_{0,h}(\mathbb{X}) = \left\{ x \in C([0, \infty); \mathbb{X}) : \lim_{t \to \infty} \frac{\| x(t) \|}{h(t)} = 0 \right\},$$

which we equip with the norm

$$\|x\|_h = \sup_{t \geq 0} \frac{\| x(t) \|}{h(t)}.$$

Recall the following well-known compactness criterion:

Lemma 10.10 *A set $B \subset C_0([0, \infty); \mathbb{X})$ is relatively compact in $C_0([0, \infty); \mathbb{X})$ if and only if,*

(a) *B is equi-continuous;*
(b) $\lim\limits_{t \to \infty} \| x(t) \| = 0$, *uniformly for $x \in B$;*
(c) *The set $B(t) = \{x(t) : x \in B\}$ is relatively compact in $\mathbb{X}$ for every $t \geq 0$.*

Proof The proof is left to the reader as an exercise.

The main existence result of this subsection can now be formulated as follows:

Theorem 10.11 *Under assumption* ($\mathbf{H_1}$), *if the following conditions hold:*

(a) *for every $t \in I$ and each $r \geq 0$ the sets*

$$\left\{ S(t) f(s, \psi) : (s, \psi) \in [0, t] \times B_r(0, \mathscr{B}) \right\}$$

and

$$g([0, t] \times B_r(0, \mathscr{B}))$$

are relatively compact in $\mathbb{X}$;
(b) *for every* $L \geq 0$,

$$\frac{1}{h(t)} \int_0^t m(s) W(Lh(s)) \, ds \to 0$$

as $t \to \infty$ *and*

$$\limsup_{r \to \infty} \frac{1}{r} \int_0^\infty \gamma(s) \frac{W((K+M)rh(s))}{h(s)} ds < 1.$$

Then the system Eqs. (10.12)–(10.14) has a mild solution on $[0, \infty)$.

Proof On the space

$$\mathscr{B}C_{0,h}(\mathbb{X}) = \{x : \mathbb{R} \to \mathbb{X} : x_0 \in \mathscr{B}, \, x|_I \in C_{0,h}(\mathbb{X})\}$$

endowed with the norm defined by

$$\|x\|_{\mathscr{B}C_{0,h}} = \|x_0\|_{\mathscr{B}} + \|x|_I\|_h,$$

we define the map $\Gamma : \mathscr{B}C_{0,h}(\mathbb{X}) \to \mathscr{B}C_{0,h}(\mathbb{X})$ by $(\Gamma x)_0 = \varphi$ and

$$\Gamma x(t) = C(t)\varphi(0) + S(t)[\xi - g(0, \varphi)] + \int_0^t C(t-s)g(s, x_s)ds$$

$$+ \int_0^t S(t-s)f(s, x_s)ds, \quad t \geq 0.$$

It is easy to prove that the expression $\Gamma x(\cdot)$ is well defined for each $x \in \mathscr{B}C_{0,h}(\mathbb{X})$. On the other hand, using the fact that $\|x_s\|_{\mathscr{B}} \leq (K+M) \| x \|_{\mathscr{B}C_{0,h}} h(s)$ for $s \in I$, we find that

$$\frac{\|\Gamma x(t)\|}{h(t)} \leq \frac{NH\|\varphi\|_{\mathscr{B}} + (\|\xi\| + \|g(0, \varphi)\|)}{h(t)} \tag{10.21}$$

$$+ \frac{1}{h(t)} \int_0^t [Nm_g(s) + \tilde{N}m_f(s)]W((K+M) \| x \|_{\mathscr{B}C_{0,h}} h(s))ds,$$

which implies, from condition **(c)**, that $\frac{\|\Gamma x(t)\|}{h(t)}$ converges to zero as $t \to \infty$. This shows that Γ is a well-defined map from $\mathscr{B}C_{0,h}(\mathbb{X})$ into $\mathscr{B}C_{0,h}(\mathbb{X})$. Note that the inequality (10.21) shows also that $\frac{\|\Gamma x(t)\|}{h(t)} \to 0$, as $t \to \infty$, uniformly for x in bounded sets of $\mathscr{B}C_{0,h}(\mathbb{X})$.

In the sequel we prove that Γ verifies the hypotheses of Theorem 1.81. We begin by proving that Γ is continuous. Let $(u^n)_n$ be a sequence in $\mathscr{B}C_{0,h}(\mathbb{X})$ and $u \in \mathscr{B}C_{0,h}(\mathbb{X})$ such that $u^n \to u$ as $n \to \infty$. Clearly, $g(s, u_s^n) \to g(s, u_s)$, $f(s, u_s^n) \to f(s, u_s)$ a.e. $s \in I$ as $n \to \infty$, and

$$\|f(s, u_s^n)\| \le m_f(s)W_f(\beta h(s)), \quad s \ge 0,$$

$$\|g(s, u_s^n)\| \le m_g(s)W_g(\beta h(s)), \quad s \ge 0,$$

where $\beta = (K + M)L$ and $L > 0$ is such that

$$\sup\{\| u \|_{\mathscr{B}C_{0,h}(\mathbb{X})}, \|u^n\|_{\mathscr{B}C_{0,h}(\mathbb{X})} : n \in \mathbb{N}\} \le L.$$

Since the functions on the right-hand side of the above inequalities (involving f and g) are integrable on $[0, t]$, we conclude that

$$\| \Gamma u^n(t) - \Gamma u(t) \| \to 0 \text{ as } n \to \infty$$

uniformly for t in bounded intervals. Moreover, using the argument that the set of functions $\{u^n : n \in \mathbb{N}\}$ is bounded in $\mathscr{B}C_{0,h}(\mathbb{X})$, for each $\epsilon > 0$ there exists $T_\epsilon > 0$ such that $\frac{\|\Gamma u^n(t) - \Gamma u(t)\|}{h(t)} \le \epsilon$, for all $n \in \mathbb{N}$ and every $t \ge T_\epsilon$. Combining these properties we obtain that $\Gamma u^n \to \Gamma u$ in $\mathscr{B}C_{0,h}(\mathbb{X})$. Thus, Γ is continuous.

On the other hand, if $x^\lambda \in \mathscr{B}C_{0,h}(\mathbb{X})$ is a solution of the integral equation $\lambda \Gamma z = z$, $0 < \lambda < 1$, for $t \ge 0$, we obtain that

$$\frac{\|x^\lambda(t)\|}{h(t)} \le \frac{NH \| \varphi \|_{\mathscr{B}} + \widetilde{N}(\| \xi \| + \|g(0, \varphi)\|)}{h(t)}$$

$$+ \frac{1}{h(t)} \int_0^t \gamma(s)W((K + M) \| x^\lambda \|_{\mathscr{B}C_{0,h}(\mathbb{X})} h(s))ds,$$

and hence

$$\| x^\lambda \|_{\mathscr{B}C_{0,h}(\mathbb{X})} \le (1 + NH) \| \varphi \|_{\mathscr{B}} + \widetilde{N}(\| \xi \| + \|g(0, \varphi)\|)$$

$$+ \int_0^\infty \gamma(s) \frac{W((K + M) \| x^\lambda \|_{\mathscr{B}C_{0,h}(\mathbb{X})} h(s))}{h(s)}ds.$$

From the previous estimates, if the set $\{\| x^\lambda \|_{\mathscr{B}C_{0,h}(\mathbb{X})} : 0 < \lambda < 1\}$ is unbounded, we deduce the existence of a sequence $(r^n)_{n \in \mathbb{N}}$ with $r^n \to \infty$ such that

$$1 \le \liminf_{n \to \infty} \frac{1}{r_n} \int_0^\infty \gamma(s) \frac{W((K + M)r_n h(s))}{h(s)}ds$$

which is absurd, therefore the set $\{\| x^\lambda \|_{\mathscr{B}C_{0,h}(\mathbb{X})} : 0 < \lambda < 1, \}$ is bounded.

Arguing as in the proof of Theorem 10.5, we can prove that

$$\left\{ \Gamma x(t) : x \in B_r(0, \mathscr{B}C_{0,h}(\mathbb{X})) \right\}$$

is relatively compact in $\mathbb{X}$ for every $t \geq 0$ and that

$$\left\{ \frac{\Gamma x}{h} : x \in B_r(0, \mathscr{B}C_{0,h}(\mathbb{X})) \right\}$$

is equi-continuous on $[0, \infty)$. Moreover, from our previous remarks we know that $\frac{\Gamma x(t)}{h(t)} \to 0$ as $t \to \infty$, uniformly for $x \in B_r(0, \mathscr{B}C_{0,h}(\mathbb{X}))$. Consequently, we have shown that the set

$$\left\{ \frac{\Gamma x}{h} : x \in B_r(0, \mathscr{B}C_{0,h}(\mathbb{X})) \right\}$$

fulfills the conditions of Lemma 10.10, which yields it is relatively compact in $C_0(\mathbb{X})$. Therefore, $\Gamma B_r(0, \mathscr{B}C_{0,h}(\mathbb{X}))$ is relatively compact in $\mathscr{B}C_{0,h}(\mathbb{X})$.

The existence of a mild solution for the system Eqs. (10.12)–(10.14) on $[0, \infty)$ follows from Theorem 1.81.

10.3.6   Existence of Asymptotically Almost Periodic Solutions

In this subsection we study the existence of asymptotically almost periodic solutions for the abstract system Eqs. (10.12)–(10.14). For that, suppose that there exist two positive constants N and $\widetilde{N}$ such that

$$\|C(t)\| \leq N$$

and

$$\|S(t)\| \leq \widetilde{N}$$

for every $t \geq 0$.

Let us recall the following definitions which are needed in the sequel:

Definition 10.12 An operator function $F : [0, \infty) \to B(\mathscr{V}, \mathscr{W})$ is said to be:

(a) strongly continuous if for every each $x \in \mathscr{V}$, the function $F(\cdot)x : [0, \infty) \to \mathscr{W}$ is continuous;

(b) pointwise almost periodic (respectively, pointwise asymptotically almost periodic) if $F(\cdot)x \in AP(\mathscr{W})$ for every $x \in \mathscr{V}$ (respectively, $F(\cdot)x \in AAP(\mathscr{W})$ for every $x \in \mathscr{V}$);

(c) almost periodic (respectively, asymptotically almost periodic) if $F(\cdot) \in AP(B(\mathcal{V}, \mathcal{W}))$ (respectively, $F(\cdot) \in AAP(B(\mathcal{V}, \mathcal{W}))$).

Remark 10.13 Note that if the sine function $S(\cdot)$ is uniformly bounded and pointwise almost periodic, then the cosine function $C(\cdot)$ is also pointwise almost periodic, see, e.g., [64, Lemma 3.1] and [64, Theorem 3.2] for details.

Lemma 10.14 ([113, Chapter 6]) *Let* $V \subseteq AP(\mathbb{X})$ *be a set with the following properties:*

(a) *V is uniformly equi-continuous on $\mathbb{R}$;*
(b) *for each $t \in \mathbb{R}$, the set $V(t) = \{x(t) : x \in V\}$ is relatively compact in $\mathbb{X}$;*
(c) *V is equi-almost periodic, that is, for every $\varepsilon > 0$ there is a relatively dense set $\mathscr{H}(\varepsilon, V, \mathbb{X}) \subset \mathbb{R}$ such that*

$$\|x(t + \tau) - x(t)\| \leq \varepsilon, \quad x \in V, \quad \tau \in \mathscr{H}(\varepsilon, V, \mathbb{X}), \, t \in \mathbb{R}.$$

Then, V is relatively compact in $AP(\mathbb{X})$.

Remark 10.15 As an immediate consequence of this characterization, one can assert that if $F : \mathbb{R} \to B(\mathbb{X}, \mathbb{Y})$ is almost periodic and U is a relatively compact subset of $\mathbb{X}$, then $V = \{F(\cdot)x : x \in U\}$ is relatively compact in $AP(\mathbb{Y})$. For the sine function, we can strengthen this property.

Proposition 10.16 *Assume that the sine function $S(\cdot)$ is almost periodic and that $U \subseteq \mathbb{X}$. If the set $\{S(t)x : x \in U, \, t \geq 0\}$ is relatively compact in $\mathbb{X}$, then $V = \{S(\cdot)x : x \in U\}$ is relatively compact in $AP(\mathbb{X})$.*

Proof Let us fix $\delta > 0$. Since $S(\delta)U$ is relatively compact in $\mathbb{X}$, by using Remark 10.15, we can claim that $V_\delta = \{S(\cdot)S(\delta)x : x \in U\}$ is relatively compact in $AP(\mathbb{X})$. On the other hand, for each $\varepsilon > 0$ there is $\delta > 0$ such that

$$\|(I - C(s))S(t)x\| \leq \varepsilon$$

for all $0 \leq s \leq \delta$, every $x \in U$ and all $t \geq 0$.
    We deduce from above that

$$\| S(t)x - \frac{1}{\delta}S(t)S(\delta)x \| = \| \frac{1}{\delta}\int_0^\delta (I - C(s))S(t)x\,ds \| \leq \varepsilon,$$

for every $t \geq 0$.
    This property and the decomposition

$$S(\cdot)x = \frac{1}{\delta}S(\cdot)S(\delta)x + S(\cdot)x - \frac{1}{\delta}S(\cdot)S(\delta)x,$$

imply that $V \subseteq \frac{1}{\delta}V_\delta + B_\varepsilon(0, C_b(\mathbb{X}))$, which in turn proves that V is relatively compact in $AP(\mathbb{X})$.

Remark 10.17 Recall that the assumption on the compactness of the set $\{S(t)x :$ $x \in U, \ t \geq 0\}$ in Proposition 10.16 is verified, for instance, in the case when the sine function is almost periodic.

Lemma 10.18 *Assume that $S(\cdot)$ is pointwise almost periodic and that U is a bounded subset of $\mathbb{X}$. If one of the following conditions holds:*

 (i) *U is relatively compact.*
(ii) *$S(\cdot)$ is almost periodic and $S(t)$ is compact for every $t \in \mathbb{R}$.*

Then $\{S(t)x : x \in U, \ t \geq 0\}$ is relatively compact in $\mathbb{X}$.

Proof The proof is left to the reader as an exercise.

Recall that the case (ii) includes the case of periodic sine functions.

For asymptotically almost periodic functions, we have a similar characterization of compactness given in the next lemma.

Lemma 10.19 *Let $V \subseteq AAP(\mathbb{X})$ be a set with the following properties:*

(a) *V is uniformly equi-continuous on $[0, \infty)$;*
(b) *for each $t \geq 0$, the set $V(t) = \{x(t) : x \in V\}$ is relatively compact in $\mathbb{X}$;*
(c) *V is equi-asymptotically almost periodic, that is, for every $\varepsilon > 0$ there are $L(\varepsilon, V, \mathbb{X}) \geq 0$ and a relatively dense set $\mathscr{H}(\varepsilon, V, \mathbb{X}) \subseteq [0, \infty)$ such that*

$$\|x(t + \tau) - x(t)\| \leq \varepsilon, \quad x \in V, \ t \geq L(\varepsilon, V, \mathbb{X}), \ \tau \in \mathscr{H}(\varepsilon, V, \mathbb{X}).$$

Then, V is relatively compact in $AAP(\mathbb{X})$.

Proof The proof is left to the reader as an exercise.

Remark 10.20 If $f \in AAP(\mathbb{X})$, then it can be decomposed in a unique fashion as $f = f_1 + f_2$, where $f_1 \in AP(\mathbb{X})$ and $f_2 \in C_0(\mathbb{X})$. Let $V \subseteq AAP(\mathbb{X})$ and $V_i = \{f_i : f \in V\}$, $i = 1, 2$. It follows from the above-mentioned results that V is relatively compact in $AAP(\mathbb{X})$ if, and only if, V_1 is relatively compact in $AP(\mathbb{X})$ and V_2 is relatively compact in $C_0(\mathbb{X})$.

We will be using the next proposition.

Proposition 10.21 *Let $(\mathscr{V}_i, \| \cdot \|_{\mathscr{V}_i})$, $i = 1, 2$, be Banach spaces and $V \subseteq L^1([0, \infty), \mathscr{V}_1)$. If $F_1 : [0, \infty) \to B(\mathscr{V}_1, \mathscr{V}_2)$ and $F_2 : [0, \infty) \to B(\mathscr{V}_2)$ are strongly continuous functions of bounded linear operators which satisfy*

(a) $\displaystyle\int_L^\infty F_1(s)x(s)ds \to 0$ *in $\mathscr{V}_2$ when $L \to \infty$, uniformly for $x \in V$;*
(b) *For each $t \geq 0$, the set $\{x(s) : x \in V, \ 0 \leq s \leq t\}$ is relatively compact in $\mathscr{V}_1$,*

then the sets

$$W(t) = \left\{ \int_0^t F_1(s)x(s)ds : x \in V \right\}, \quad t \geq 0,$$

and

$$W = \bigcup_{0 \le t \le \infty} W(t)$$

are relatively compact in $\mathcal{V}_2$. Moreover, if F_2 is uniformly bounded on $[0, \infty)$ and

$$\int_t^{t+h} F_1(s)x(s)ds \to 0,$$

as $h \to 0$, uniformly for $x \in V$, then the set $U = \{z_x : x \in V\}$, where

$$z_x(t) = F_2(t) \int_t^\infty F_1(s)x(s)ds,$$

is relatively compact in $C_0(\mathcal{V}_2)$.

Proof Let $(K_t)_{t \ge 0}$ be a family of compacts sets such that $\{x(s) : x \in V, s \in [0, t]\} \subseteq K_t$ for every $t \ge 0$. Since F_1 is strongly continuous, then the set

$$F_1 K_t = \{F_1(s)y : y \in K_t, 0 \le s \le t\}$$

is relatively compact in $\mathcal{V}_2$.

Let $(\widetilde{K}_t)_{t \ge 0}$ be a nondecreasing family of compact and absolutely convex sets such that $F_1 K_t \subset \widetilde{K}_t$ for every $t \ge 0$.

From the mean value theorem for the Bochner integral (see [90, Lemma 2.1.3]), we infer that $W(t) \subseteq t\widetilde{K}_t$ for all $t > 0$. On the other hand, for each $\varepsilon > 0$ there is a constant $L \ge 0$ such that

$$\left\| \int_L^\infty F_1(s)x(s)ds \right\|_{\mathcal{V}_2} \le \varepsilon$$

for all $x \in V$.

Using the sets $\widetilde{K}_t$ it follows that $W \subseteq L\widetilde{K}_L + B_\varepsilon(0, \mathcal{V}_2)$, which yields W is relatively compact in $\mathcal{V}_2$. Thus, the sets $W(t), t \ge 0$, and W are relatively compact in $\mathcal{V}_2$.

To establish the last assertion, we make use of Lemma 10.10. The hypothesis (**b**) of Lemma 10.10 can be easily obtained as an immediate consequence of (**a**) and the fact that F_2 is uniformly bounded. Moreover, for every $t \ge 0$ and $x \in V$, we have that

$$\int_t^\infty F_1(s)x(s)ds \in \overline{W - W(t)} \subset \overline{W - W} = W_1,$$

which proves that the set

$$U(t) = \left\{ F_2(t) \int_t^\infty F_1(s)x(s)ds : x \in V \right\}$$

is relatively compact in $\mathcal{V}_2$ for every $t \geq 0$.

Finally, we prove that U is equi-continuous. To this end, we fix $t \geq 0$. Since elements

$$\int_t^\infty F_1(\xi)x(\xi)\,d\xi, \quad x \in V,$$

are in the compact set W_1 (which is independent of t), and the family $(F_2(t))_{t \geq 0}$ is strongly continuous in $\mathcal{V}_2$, for $\varepsilon > 0$ there exists $\delta > 0$ such that

$$\| F_2(t+s)x - F_2(t)x \| \leq \varepsilon, \quad x \subset W_1,$$

$$\left\| \int_t^{t+s} F_1(\xi)x(\xi)d\xi \right\| \leq \varepsilon, \quad x \in V,$$

for every $0 <| s |< \delta$ with $t + s \geq 0$. Consequently, for $x \in V$ and $0 <| s |< \delta$ such that $t + s \geq 0$, we get,

$$\left\| F_2(t+s) \int_{t+s}^\infty F_1(\xi)x(\xi)\,d\xi - F_2(t) \int_t^\infty F_1(\xi)x(\xi)\,d\xi \right\|$$

$$\leq \left\| (F_2(t+s) - F_2(t)) \int_{t+s}^\infty F_1(\xi)x(\xi)d\xi \right\|$$

$$+ \| F_2(t) \| \left\| \int_{t \wedge (t+s)}^{t \vee (t+s)} F_1(\xi)x(\xi)\,d\xi \right\|$$

$$\leq \sup\{\| (F_2(t+s)y - F_2(t)y \|: y \in W_1\} + \| F_2(t) \| \varepsilon,$$

$$\leq (1 + \sup_{\theta \geq 0} \| F_2(\theta) \|)\varepsilon,$$

which implies that U is equi-continuous at t.

In the next results, for a locally integrable function $x : [0, \infty) \to X$, we denote by $z_x, y_x : [0, \infty) \to \mathbb{X}$ the functions given by

$$z_x(t) = \int_0^t C(t - s)x(s)ds$$

and

$$y_x(t) = \int_0^t S(t - s)x(s)ds.$$

Proposition 10.22 *Assume that $S(\cdot)$ is pointwise almost periodic and that $V \subseteq L^1([0, \infty), \mathbb{X})$ is a set with the following properties:*

(a) $\displaystyle\int_L^\infty \|x(s)\| ds \to 0$ *when $L \to \infty$, uniformly for $x \in V$;*

(b) $\displaystyle\int_t^{t+s} \|x(\xi)\| d\xi \to 0$, *when $s \to 0$, uniformly for $x \in V$ and $t \geq 0$;*

(c) *for each $t \geq 0$ the set $\{x(s) : 0 \leq s \leq t, \; x \in V\}$ is relatively compact.*

Then the sets $\{y_x : x \in V\}$ and $\{z_x : x \in V\}$ are relatively compact in $AAP(\mathbb{X})$.

Proof We first establish that each function y_x is asymptotically almost periodic. For $x \in V$, we can write

$$y_x(t) = S(t)\int_0^t C(s)x(s)\,ds \; - \; C(t)\int_0^t S(s)x(s)\,ds$$

$$= S(t)\int_0^\infty C(s)x(s)\,ds \; - \; S(t)\int_t^\infty C(s)x(s)\,ds$$

$$- \; C(t)\int_0^\infty S(s)x(s)\,ds \; + \; C(t)\int_t^\infty S(s)x(s)\,ds.$$

Since the sine function $S(\cdot)$ is pointwise almost periodic, it follows from [64, Lemma 3.1] and [64, Theorem 3.2] that $C(\cdot)$ is also pointwise almost periodic. Therefore, the first and third terms on the right-hand side define almost periodic functions while the second and fourth terms are functions that vanish at ∞. Thus, $y_x \in AAP(\mathbb{X})$.

From Proposition 10.21, we know that the integrals

$$\int_0^\infty C(s)x(s)ds$$

and

$$\int_0^\infty S(s)x(s)ds,$$

for $x \in V$, are included in a compact subset of $\mathbb{X}$, which implies that the set formed by the functions

$$S(\cdot)\int_0^\infty C(s)x(s)ds - C(\cdot)\int_0^\infty S(s)x(s)ds, \; x \in V,$$

is relatively compact in $AP(\mathbb{X})$. The same Proposition enables us to infer that the set

$$\{t \to C(t) \int_t^\infty S(s)x(s)ds - S(t) \int_t^\infty C(s)x(s)ds : x \in V\}$$

is relatively compact in $C_0(\mathbb{X})$. This shows that $\{y_x : x \in V\}$ is relatively compact in $AAP(\mathbb{X})$.

We now prove that the set $\{z_x : x \in V\}$ is relatively compact in $AAP(\mathbb{X})$. For that we first show that the functions z_x, $x \in V$, are uniformly continuous. First of all, fix $L > 0$. Since $C(\cdot)$ is pointwise almost periodic, from (c) we have that

$$\|(C(t+s) - C(t))x(\xi)\| \to 0,$$

as $s \to 0$, uniformly for $t \geq 0$, $0 \leq \xi \leq L$ and $x \in V$. Therefore,

$$\begin{aligned}
\|z_x(t+s) &- z_x(t)\| \\
&\leq \int_0^{t \wedge (t+s)} \|C(t+s-\xi)x(\xi) - C(t-\xi)x(\xi)\|d\xi \\
&\quad + \|\int_{t \wedge (t+s)}^{t \vee (t+s)} C(t+s-\xi)x(\xi)d\xi\| \\
&\leq \int_0^L \sup_{t \geq 0, x \in V} \|(C(t+s-\xi) - C(t-\xi))x(\xi)\|d\xi \\
&\quad + 2N \int_L^\infty \|x(\xi)\|d\xi + N \int_{t \wedge (t+s)}^{t \vee (t+s)} \|x(\xi)\|d\xi.
\end{aligned}$$

Using conditions (a) and (b) we can appropriately choose L to show that the right-hand side of the above inequality converges to 0 as $s \to 0$, uniformly in $t \geq 0$ and $x \in V$, which proves that each function z_x is uniformly continuous on $[0, \infty)$. Moreover, from the above, it is clear that the set $\{z_x : x \in V\}$ is uniformly equi-continuous on $[0, \infty)$. Since z_x is the derivative of y_x, it follows from [113, Theorem 5.2] that $\{z_x : x \in V\}$ is a uniformly equi-continuous subset of $AAP(\mathbb{X})$. Moreover, from Proposition 10.21 we obtain that $\{z_x(t) : x \in V\}$ is relatively compact, for all $t \geq 0$.

Finally, we establish that $\{z_x : x \in V\}$ is equi-asymptotically almost periodic. For a given $\varepsilon > 0$, there exists $L_\varepsilon > 0$ such that

$$\int_{L_\varepsilon}^\infty \|x(s)\|ds \leq \varepsilon/6N$$

for all $x \in V$.

In addition, since the set $\{C(\cdot)x(s) : 0 \leq s \leq L_\varepsilon\}$ is equi-almost periodic, there is a relatively dense set $P_\varepsilon \subseteq [0, \infty)$ such that

$$\|C(\xi + \tau)x(s) - C(\xi)x(s)\| \leq \frac{\varepsilon}{3L_\varepsilon},$$

for all $\xi \geq 0$, $0 \leq s \leq L_\varepsilon$ and every $\tau \in P_\varepsilon$. Hence, for $t \geq L_\varepsilon$ and $\tau \in P_\varepsilon$, we obtain

$$\|z_x(t + \tau) - z_x(t)\| \leq \int_0^t \|C(t + \tau - s)x(s) - C(t - s)x(s)\| ds$$

$$+ \int_t^{t+\tau} \|C(t + \tau - s)x(s)\| \, ds$$

$$\leq \int_0^{L_\varepsilon} \|C(t + \tau - s)x(s) - C(t - s)x(s)\| ds$$

$$+ 3N \int_{L_\varepsilon}^\infty \|x(s)\| \, ds$$

$$\leq \varepsilon$$

which shows the assertion.

One completes the proof by applying Lemma 10.19 to the set $\{z_x : x \in V\}$.

Using this result and proceeding as in the proof of Proposition 10.16 we obtain the compactness of $\{y_x : x \in V\}$ with some weaker conditions.

Proposition 10.23 *Assume that $S(\cdot)$ is almost periodic and that $V \subseteq L^1([0, \infty), \mathbb{X})$ is uniformly bounded and satisfies the following properties:*

(a) $\displaystyle\int_L^\infty \| x(s) \| \, ds \to 0$, *when $L \to \infty$, uniformly for $x \in V$;*

(b) $\displaystyle\int_t^{t+s} \|x(\xi)\| \, d\xi \to 0$, *as $s \to 0$, uniformly for $t \geq 0$ and $x \in V$;*

(c) *for each t, $\delta \geq 0$, the set $\{S(\delta)x(s) : 0 \leq s \leq t, \, x \in V\}$ is relatively compact in $\mathbb{X}$.*

Then $\{y_x : x \in V\}$ is relatively compact in $AAP(\mathbb{X})$.

Proof Define for all $x \in V$, the function

$$\widetilde{y}_x(t) = \int_0^t S(s)x(s)ds.$$

Let $0 < \varepsilon < t \leq a$. Since the function $s \to S(s)$ is Lipschitz continuous, we can choose points $0 = t_1 < t_2 \ldots < t_n = t$ such that $\| S(s) - S(s') \| \leq \varepsilon$ for $s, s' \in [t_i, t_{i+1}]$ and $i = 1, 2, \ldots . n - 1$. For $x \in V$, then from the Mean Value Theorem for the Bochner integral (see [90, Lemma 2.1.3]), we find that

$$\widetilde{y}_x(t) = \sum_{i=1}^{n-1} \int_{t_i}^{t_{i+1}} (S(s) - S(t_i))x(s)ds + \sum_{i=1}^{n-1} \int_{t_i}^{t_{i+1}} S(t_i)x(s)ds$$

$$\in \mathscr{C}_\varepsilon + \sum_{i=1}^{n-1} (t_{i+1} - t_i) \overline{co(\{S(t_i)z(s) : s \in [0, t_i], z \in V\})}$$

$$\subset \mathscr{C}_\varepsilon + \mathscr{K}_\varepsilon,$$

where $\mathscr{K}_\varepsilon$ is compact and $diam(\mathscr{C}_\varepsilon) \to 0$ as $\varepsilon \to 0$. This proves that $W_1(t) = \{\widetilde{y}_x(t); \ x \in V\}$ is relatively compact in $\mathbb{X}$. Moreover, proceeding as in the proof of Proposition 10.21, we infer that

$$W = \bigcup_{0 \leq t \leq \infty} W(t)$$

and

$$U = \left\{ \int_t^\infty S(s)x(s)ds : x \in V, \ t \geq 0 \right\}$$

are also relatively compact in $\mathbb{X}$.

To complete the proof, we consider one more time the decomposition

$$y_x(t) = S(t) \int_0^\infty C(s)x(s)ds - S(t) \int_t^\infty C(s)x(s)ds$$

$$-C(t) \int_0^\infty S(s)x(s)ds + C(t) \int_t^\infty S(s)x(s)ds.$$

Since the cosine function is pointwise almost periodic (see Remark 10.13), we infer from Remark 10.15 and Lemma 10.10 that the set of functions

$$\left\{ t \to -C(t) \int_0^\infty S(s)x(s)\,ds + C(t) \int_t^\infty S(s)x(s) \right\}$$

is relatively compact in $AAP(\mathbb{X})$. Moreover, using the fact that $S(\cdot)$ is almost periodic and that $S(t)$ is a compact operator for every $t \geq 0$, we can prove from Remark 10.15 and Lemma 10.10 that the set of functions

$$\left\{ t \to S(t) \int_0^\infty C(s)x(s)ds - S(t) \int_t^\infty C(s)x(s)ds : x \in V \right\}$$

is also completely continuous in $AAP(\mathbb{X})$.

The main result of this subsection can be formulated as follows:

Theorem 10.24 *Assume that $S(\cdot)$ is almost periodic and that condition* $(\mathbf{H_1})$ *holds with $m_f(\cdot)$ and $m_g(\cdot)$ in $L^1([0,\infty))$. Suppose, in addition, that for every $t \geq 0$ and each $r \geq 0$ the sets $\{S(t)f(s,\psi) : (s,\psi) \in [0,t] \times B_r(0,\mathscr{B})\}$ and $g([0,t] \times B_r(0,\mathscr{B}))$ are relatively compact in $\mathbb{X}$. If*

$$K \int_0^\infty (Nm_g(s) + \widetilde{N}m_f(s))ds < \int_c^\infty \frac{ds}{W(s)}, \tag{10.22}$$

where $c = (KNH + M) \parallel \varphi \parallel_{\mathscr{B}} + K\widetilde{N}(\parallel \xi \parallel + \parallel g(0,\varphi) \parallel)$, then there exists a mild solution $u(\cdot) \in AAP(\mathscr{B}, \mathbb{X})$ to the system Eqs. (10.12)–(10.14).

Proof Let $\mathscr{B}AAP = \{x : \mathbb{R} \to \mathbb{X} : x_0 \in \mathscr{B}, x \mid_{[0,\infty)} \in AAP(\mathbb{X})\}$ endowed with the semi-norm $\parallel x \parallel_{\mathscr{B}AAP} := \parallel x_0 \parallel_{\mathscr{B}} + \sup_{t \geq 0} \parallel x(t) \parallel$ and $\Gamma : \mathscr{B}AAP \to \mathscr{B}AAP$ be the operator defined by

$$\Gamma x(t) = C(t)\varphi(0) + S(t)[\xi - g(0,\varphi)]$$

$$+ \int_0^t C(t-s)g(s,x_s)ds + \int_0^t S(t-s)f(s,x_s)ds,$$

for $t \geq 0$, and $(\Gamma x)_0 = \varphi$.

By the integrability of the functions $m_f(\cdot)$ and $m_g(\cdot)$ and proceeding as in the proof of Proposition 10.22 for the functions $f(s,x_s)$ and $g(s,x_s)$, we infer that $\Gamma(x) \in AAP(\mathscr{B}, \mathbb{X})$. Furthermore, if we take a sequence $(x^n)_n$ that converges to x in the space $AAP(\mathscr{B}, \mathbb{X})$, then $S(t-s)f(s,x_s^n) \to S(t-s)f(s,x_s)$ and $C(t-s)g(s,x_s^n) \to C(t-s)g(s,x_s)$, as $n \to \infty$, a.e. for $t,s \in [0,\infty]$. Let $L = \sup\{\parallel x \parallel_{\mathscr{B}C}, \parallel x^n \parallel_{\mathscr{B}C} : n \in \mathbb{N}\}$ and $\beta = (K+M)L$. From the inequalities

$$\parallel C(t-s)g(s,x_s^n) - C(t-s)g(s,x_s) \parallel \leq 2Nm_g(s)W_g(\beta),$$

$$\parallel S(t-s)f(s,x_s^n) - S(t-s)f(s,x_s) \parallel \leq 2\widetilde{N}m_f(s)W_f(\beta),$$

and using the integrability of $m_f(\cdot)$ and $m_g(\cdot)$, we conclude that $\parallel \Gamma x^n - \Gamma x \parallel_{\mathscr{B}AAP} \to 0$ when $n \to \infty$. Thus, Γ is a continuous map from $AAP(\mathscr{B}, \mathbb{X})$ into $AAP(\mathscr{B}, \mathbb{X})$.

On the other hand, proceeding as in the proof of Theorem 10.5, we conclude that the set of functions $\{x^\lambda \in AAP(\mathscr{B}, \mathbb{X}) : \lambda\Gamma(x^\lambda) = x^\lambda, 0 < \lambda < 1\}$ is uniformly bounded on $[0,\infty)$.

Finally, we show that Γ is completely continuous. In order to establish this assertion, we take a bounded set $V \subseteq AAP(\mathscr{B}, \mathbb{X})$. Since the sets of functions $\Lambda_1 = \{s \to g(s,x_s) : x \in V\}$ and $\Lambda_2 = \{s \to f(s,x_s) : x \in V\}$ satisfy the hypotheses of Propositions 10.22 and 10.23, respectively, we deduce that $\Gamma(V)$ is relatively compact in $AAP(\mathbb{X})$. The assertion is now a consequence of Theorem 1.81.

In Theorem 10.26 below, we prove the existence of an asymptotically almost periodic mild solution to Eqs. (10.12)–(10.14) assuming that $g(\cdot)$ satisfies an appropriate Lipschitz condition. For that, we need the following lemma.

Lemma 10.25 *If $\mathscr{B}$ is a fading memory space and $z \in BC(\mathbb{R}; \mathbb{X})$ is a function such that $z_0 \in \mathscr{B}$ and $z \in AAP(\mathbb{X})$, then $t \to z_t \in AAP(\mathscr{B})$.*

Theorem 10.26 *Assume that the sine function $S(\cdot)$ is almost periodic and that $\mathscr{B}$ is a fading memory space. Suppose, in addition, that the following conditions hold:*

(a) *For every $t \geq 0$ and each $r \geq 0$, the set*

$$\left\{ S(t)f(s, \psi) : (s, \psi) \in [0, t] \times B_r(0, \mathscr{B}) \right\}$$

is relatively compact in $\mathbb{X}$.
(b) *There exists a function $L_g \in L^1([0, \infty))$ such that*

$$\| g(t, \psi_1) - g(t, \psi_2) \| \leq L_g(t) \| \psi_1 - \psi_2 \|_{\mathscr{B}}, \quad (t, \psi_j) \in [0, \infty) \times \mathscr{B}.$$

(c) *The condition $(\mathbf{H_1})$ is valid with m_g, m_f in $L^1([0, \infty))$ and*

$$(K + M) \liminf_{\xi \to \infty} \frac{W(\xi)}{\xi} \int_0^\infty (Nm_g(s) + \tilde{N}m_f(s))ds < 1. \qquad (10.23)$$

Then there exists a mild solution $u(\cdot) \in AAP(\mathbb{X})$ of (10.12)–(10.14).

Proof Let $\mathscr{B}AAP = \{x : \mathbb{R} \to \mathbb{X} : x_0 \in \mathscr{B}, x \mid_{[0,\infty)} \in AAP(\mathbb{X})\}$ endowed with the semi-norm defined by $\|x\|_{\mathscr{B}AAP} = \|x_0\|_{\mathscr{B}} + \sup_{t \geq 0} \|x(t)\|$. On this space, we define the operators $\Gamma_i : \mathscr{B}AAP \to \mathscr{B}AAP, i = 1, 2$, by

$$\Gamma_1 x(t) = C(t)\varphi(0) + S(t)[\xi - g(0, \varphi)] + \int_0^t C(t - s)g(s, x_s)ds,$$

$$\Gamma_2 x(t) = \int_0^t S(t - s)f(s, x_s)ds,$$

for $t \geq 0$, and $(\Gamma_1 x)_0 = \varphi$ and $(\Gamma_2 x)_0 = 0$.

From the proof of Proposition 10.22, we infer that the functions

$$\zeta(t) = \int_0^t S(t - s)g(s, x_s)ds$$

and $\Gamma_2 x$ are asymptotically almost periodic. It is easy to see that

$$\zeta'(t) = \int_0^t C(t - s)g(s, x_s)ds.$$

Moreover, since the function $s \to g(s, x_s)$ is integrable on $[0, \infty)$ and

$$\|\zeta'(t + h) - \zeta'(t)\| \leq \int_0^h N \parallel g(s, x_s) \parallel ds$$

$$+N \int_0^\infty \parallel g(s + h, x_{s+h}) - g(s, x_s) \parallel ds,$$

converge to zero as $h \to 0$, uniformly for $t \in [0, \infty)$, we can conclude from [113, Theorem 5.2] that $\Gamma_1 x$ is also asymptotically almost periodic. This proof that $\Gamma_1 x, \Gamma_2 x$ are well defined and that Γ_1, Γ_2 are functions defined from $\mathscr{B}AAP$ into $\mathscr{B}AAP$.

Let $y : \mathbb{R} \to \mathbb{X}$ be the extension of φ to $\mathbb{R}$ such that

$$y(t) = C(t)\varphi(0) + S(t)[\xi - g(0, \varphi)]$$

for $t \geq 0$ and $\Gamma : \mathscr{B}AAP \to \mathscr{B}AAP$ be the map $\Gamma = \Gamma_1 + \Gamma_2$. We next prove that there exists $r > 0$ such that $\Gamma(B_r(y, \mathscr{B}AAP)) \subset B_r(y, \mathscr{B}AAP)$. Proceeding by contradiction, we suppose that for each $r > 0$ there exist $u^r \in B_r(y, \mathscr{B}AAP)$ and $t^r \geq 0$ such that $\parallel \Gamma u^r(t^r) - y(t^r) \parallel > r$. Consequently,

$$r \leq \parallel \Gamma u^r(t^r) - y(t^r) \parallel$$

$$\leq \int_0^{t^r} (Nm_g(s) + \tilde{N}m_f(s))W(\parallel u_s^r - y_s \parallel_{\mathscr{B}} + \parallel y_s \parallel_{\mathscr{B}})ds$$

$$\leq \int_0^\infty (Nm_g(s) + \tilde{N}m_f(s))W((K + M)r + \rho)ds,$$

where $\rho = (M + KNH) \parallel \varphi \parallel_{\mathscr{B}} + K\tilde{N} \parallel \xi - g(0, \varphi) \parallel$, which yields

$$1 \leq (K + M) \liminf_{\xi \to \infty} \frac{W(\xi)}{\xi} \int_0^\infty (Nm_g(s) + \tilde{N}m_f(s))ds.$$

Since this inequality contradicts Eq. (10.23), we obtain the assertion.

Let $r > 0$ such that $\Gamma(B_r(0, \mathscr{B}AAP)) \subset B_r(0, \mathscr{B}AAP)$. Proceeding as in the proof of Theorem 10.24, we can show that the map Γ_2 is completely continuous. Moreover, from the estimate

$$\parallel \Gamma_1 u(t) - \Gamma_1 v(t) \parallel \leq NK \int_0^t L_g(s)ds \parallel u - v \parallel_{\mathscr{B}AAP},$$

we infer that Γ_1 is a contraction on $\mathscr{B}AAP$, which enables us to conclude that Γ is condensing on $B_r(0, \mathscr{B}AAP)$. Now, the assertion is a consequence of Theorem 1.80.

10.3.7 Asymptotically Almost Periodic Solutions to Some Second-Order Integro-differential Systems

This subsection is devoted an illustrative example to the previous subsection and consists of studying the existence of asymptotically almost periodic mild solutions for the second-order partial differential equations given by

$$\frac{\partial}{\partial t}\left[\frac{\partial u(t,\xi)}{\partial t} + \eta(t)u(t,\xi) + \int_{-\infty}^{t} \alpha_1(t,s)u(s,\xi)ds\right] = \frac{\partial^2 u(t,\xi)}{\partial \xi^2}$$

$$+ \int_{-\infty}^{t} \alpha_2(t,s)u(s,\xi)ds, \tag{10.24}$$

for $t \geq 0$ and $\xi \in J = [0, \pi]$, subject to the initial conditions

$$u(t,0) = u(t,\pi) = 0, \qquad t \geq 0, \tag{10.25}$$

$$u(\theta,\xi) = \varphi(\theta,\xi), \qquad \theta \leq 0, \ \xi \in J, \tag{10.26}$$

$$\frac{\partial u(0,\xi)}{\partial t} = z(\xi), \qquad \xi \in J, \tag{10.27}$$

where $\eta(\cdot) : \mathbb{R} \to \mathbb{R}, \alpha_i : \mathbb{R}^2 \to \mathbb{R}$ $(i = 1, 2)$ are continuous functions and φ, ξ are some appropriate functions.

In order to cast the above system into an abstract version of the previous subsection, we let $\mathbb{X} = (L^2(0, \pi); \| \cdot \|_2)$ and consider the operator $A : D(A) \subset \mathbb{X} \to \mathbb{X}$ defined by

$$D(A) = \left\{u \in H^2(0, \pi) : u(0) = u(\pi) = 0\right\}, \qquad Au = \frac{d^2 u}{dx^2}, \ u \in D(A).$$

It is well known that A is the infinitesimal generator of a strongly continuous cosine function, $(C(t))_{t \in \mathbb{R}}$ on $\mathbb{X}$. Furthermore, A has discrete spectrum with eigenvalues $-n^2$, $n \in \mathbb{N}$, with corresponding normalized eigenvectors given by

$$z_n(\xi) = \left(\frac{2}{\pi}\right)^{1/2} \sin(n\xi).$$

Moreover, the following properties are fulfilled:

(a) The set $\{z_n : n \in \mathbb{N}\}$ is an orthonormal basis of $\mathbb{X}$;

(b) For $u \in \mathbb{X}$, $C(t)u = \displaystyle\sum_{n=1}^{\infty} \cos(nt)\langle u, z_n \rangle z_n$. It follows from this expression that

$$S(t)u = \sum_{n=1}^{\infty} \frac{\sin(nt)}{n} \langle u, z_n \rangle z_n.$$

Moreover, the sine function $S(\cdot)$ is periodic with $S(t)$ being a compact operator for all $t \in \mathbb{R}$ such that $\max\{\|C(t)\|, \|S(t)\|\} \leq 1$, for every $t \in \mathbb{R}$.

As a phase space we choose the space $\mathcal{B} = C_r \times L^p(\rho; \mathbb{X})$, $r \geq 0$, $1 \leq p < \infty$ (see Example 10.2) and assume that the conditions (g-5)–(g-7) in the terminology of [69] are valid. Note that under these conditions, the space $\mathcal{B}$ is a fading memory space and that there exists $\mathfrak{K} > 0$ such that $\max\{K(t), M(t)\} \leq \mathfrak{K}$ for all $t \geq 0$, see [69, Example 7.1.8] and [69, Proposition 7.1.5] for details.

By assuming that

$$L_g(t) = |\eta(t)| + \left(\int_{-\infty}^{0} \left[\frac{\alpha_1(t, t+\theta)}{\rho(\theta)} \right]^2 d\theta \right)^{1/2},$$

$$m_f(t) = \left(\int_{-\infty}^{0} \left[\frac{\alpha_2(t, t+\theta)}{\rho(\theta)} \right]^2 d\theta \right)^{1/2},$$

are finite, for every $t \geq 0$, we can define the operators $g, f : \mathbb{R}_+ \times \mathcal{B} \to \mathbb{X}$ by the mean of the expressions

$$g(t, \psi)(\xi) = \eta(t)\psi(0, \xi) + \int_{-\infty}^{0} \alpha_1(t, t+s)\psi(s, \xi)ds,$$

$$f(t, \psi)(\xi) = \int_{-\infty}^{0} \alpha_2(t, t+s)\psi(s, \xi)ds.$$

It is easy to see that $g(t, \cdot)$ and $f(t, \cdot)$ are bounded linear operators, as $\| g(t, \cdot) \|_{B(\mathcal{B},\mathbb{X})} \leq L_g(t)$ and $\| f(t, \cdot) \|_{B(\mathcal{B},\mathbb{X})} \leq m_f(t)$ for every $t \geq 0$. The next results are a direct consequence of Theorem 10.26. Thus the details of the proof will be omitted.

Proposition 10.27 *Assume $\varphi \in \mathcal{B}$, $\eta \in \mathbb{X}$ and that $L_g(\cdot), m_f(\cdot)$ are functions in $L^1([0, \infty))$. If*

$$2\mathfrak{K} \int_0^{\infty} (L_g(s) + m_f(s))ds < 1, \tag{10.28}$$

then there exists an asymptotically almost periodic mild solution to (10.24)–(10.27).

To complete this subsection, we study the existence of asymptotically almost periodic solutions for the system (10.16). To simplify the description and for sake of brevity, we consider the case when $d = 1$. Assume that the functions $\beta(\cdot)$ and $\dfrac{\partial \beta(\cdot)}{\partial s}$ are continuous and that the expression

$$
L_g(t) = \left| \frac{\beta(t,0)}{c} \right| + \frac{1}{|c|} \left(\int_{-\infty}^{0} \left[\frac{\partial \beta(t,s)}{\partial s} \rho^{-1}(s) \right]^2 ds \right)^{1/2}
$$

defines a function in $L^1([0,\infty))$. By assuming that the solution $u(\cdot)$ of (10.16) is known on $[0,\infty)$, and defining the function $g : \mathbb{R} \times \mathcal{B} \to \mathbb{X}$ by

$$
g(t,\psi)(\xi) = \frac{\beta(t,0)}{c} \psi(0,\xi) + \frac{1}{c} \int_{0}^{\infty} \frac{\partial \beta(t,s)}{\partial s} \psi(-s,\xi) ds,
$$

we can transform system (10.16) into the abstract system (10.12)–(10.14).

Corollary 10.28 *For every $\varphi \in \mathcal{B}$ and $\xi \in \mathbb{X}$, there exists an asymptotically almost periodic mild solution of Eq. (10.16) with $u_0 = \varphi$.*

Proof This result is a particular case of Proposition 10.27. We only observe that the inequality (10.23) is automatically satisfied, as $m_f = 0$. $\quad\blacksquare$

10.4 Exercises

1. Consider the functions f_1, $f_2 : \mathbb{R} \times H_0^1(\Omega) \times H_0^1(\Omega) \to L^2(\Omega)$ are defined by

$$
f_i(t,u,\theta)(x) = f_i(t, \nabla u(x), \nabla \theta(x)) = \frac{K d_i(t)}{1 + |\nabla u(x)| + |\nabla \theta(x)|}
$$

for $x \in \Omega$, $t \in \mathbb{R}$, $i = 1, 2$, where $d_i : \mathbb{R} \mapsto \mathbb{R}$ are almost periodic functions.

a. Show that the functions f_i ($i = 1, 2$) are jointly continuous.
b. Show that f_i ($i = 1, 2$) are globally Lipschitz functions, that is, there exists $L > 0$ such that

$$
\| f_i(t,u,\theta) - f_i(t,v,\eta) \|_{L^2(\Omega)} \le L \left(\| u - v \|_{H_0^1(\Omega)}^2 + \| \theta - \eta \|_{H_0^1(\Omega)}^2 \right)^{\frac{1}{2}}
$$

for all $t \in \mathbb{R}$, u, v, η and $\theta \in H_0^1(\Omega)$.

2. Prove Corollary 10.6.
3. Prove Theorem 10.8.
4. Prove Lemma 10.10.

5. Prove Lemma 10.18.
6. Prove Lemma 10.19.
7. Prove Lemma 10.25.
8. Prove Proposition 10.27.

10.5 Comments

The existence results of Sect. 10.2 are based upon Baroun et al. [20] and Baroun [29]. For additional reading on thermoelastic systems, we refer the reader to [14, 18, 24, 44, 65, 79], and [92].

The existence results of Sect. 10.2 are based upon Diagana et al. [50]. For additional reading on the topics discussed in this section, we refer the reader to Fattorini [57] and Travis and Webb [106, 107].

Appendix
List of Abbreviations and Notations

The following notations and abbreviations will be used in the rest of the book without further explanation.

$\mathbb{R}$—field of real numbers

$\mathbb{C}$—field of complex numbers

$(\mathbb{X}, \| \cdot \|)$, $(\mathbb{X}, \| \cdot \|_1)$, $(\mathbb{Y}, \| \cdot \|_2)$, $(\mathcal{V}, \| \cdot \|_\mathcal{V})$, and $(\mathcal{W}, \| \cdot \|_\mathcal{W})$—Banach spaces

$\mathcal{H}$—Denotes a generic Hilbert space over $\mathbb{F}$

$\mathbb{N}$—Denotes the set of positive integers

$\mathbb{Z}$—Denotes the set of integers

$\mathbb{Q}$—Denotes the field of rational numbers

$\mathbb{R}$—Denotes the field of real numbers

$\mathbb{R}^n$—The n-dimensional real numbers

$\mathbb{C}$—Denotes the field of complex numbers

$\mathbb{C}^n$—Denotes the n-dimensional complex numbers

$\mathbb{S}^1 = \left\{ z \in \mathbb{C} : |z| = 1 \right\}$

$\mathbb{Z}_+ = \mathbb{N} \cup \{0\}$

$\mathbb{R}_+ = [0, \infty)$

$\| \cdot \|$—Denotes the Euclidean norm

$| \cdot |$—Denotes the absolute value on $\mathbb{F}$

$B_r(x, \mathcal{V})$—Denotes the closed ball in $\mathcal{V}$, centered at x with radius r

$\ell^p(\mathbb{N}) = \ell^p(\mathbb{N}, \mathbb{R}^N) = \left\{ x = (x(t))_{t \in \mathbb{N}} : x(t) \in \mathbb{R}^N \text{ for all } t \in \mathbb{N}, \right.$
$\left. \sum_{t=1}^{\infty} \|x(t)\|^p < \infty \right\}$

$\ell^\infty(\mathbb{N}) = \ell^\infty(\mathbb{N}, \mathbb{R}^N) = \left\{ x = (x(t))_{t \in \mathbb{N}} : x(t) \in \mathbb{R}^N, \ \exists M \geq 0, \ \|x(t)\| \leq M, \right.$
$\left. \forall t \in \mathbb{N} \right\}$

(M, d)—Denotes a metric space

$[x] = x + \mathbb{L} = \{x + \ell : \ell \in \mathbb{L}\}$—denotes a coset

$\mathbb{X} \setminus \mathbb{L} = \left\{ [x] : x \in \mathbb{X} \right\}$—Denotes a quotient space

© Springer Nature Switzerland AG 2018

T. Diagana, *Semilinear Evolution Equations and Their Applications*,
https://doi.org/10.1007/978-3-030-00449-1

$(\Omega, \mathscr{F}, \mu)$—Denotes a measure space

$$L^p(\Omega, \mu) = \left\{ f : \Omega \mapsto \mathbb{C} \text{ measurable} : \|f\|_{L^p} = \|f\|_p := \left[\int_\Omega |f(x)|^p d\mu \right]^{\frac{1}{p}} < \infty \right\}$$

$L^p(\Omega, \mu) = L^p(\Omega)$

$\|\cdot\|_{L^p} = \|\cdot\|_p$

$$L^\infty(\Omega, \mu) = \left\{ f : \Omega \mapsto \mathbb{C} \text{ measurable}, \ \exists C \geq 0 \text{ such that } |f(x)| \leq C \text{ a.e. } x \in \Omega \right\}$$

$L^\infty(\Omega, \mu) = L^\infty(\Omega)$

$$D^\alpha = \frac{\partial^{|\alpha|}}{\partial x_1^{\alpha_1} \partial x_2^{\alpha_2} ... \partial x_n^{\alpha_n}}$$

$\alpha = (\alpha_1, \alpha_2, ..., \alpha_n), \alpha_i \in \mathbb{N}$ for $i = 1, ..., n$, and $|\alpha| = \alpha_1 + \alpha_2 + ... + \alpha_n$

$$W^{k,p}(\Omega) := \left\{ u \in L^p(\Omega) : D^\alpha u \in L^p(\Omega) \text{ for } |\alpha| \leq k \right\}$$

$W^{k,2}(\Omega) = H^k(\Omega)$

$$\|u\|_{k,p} = \|u\|_{W^{k,p}} = \left(\sum_{|\alpha| \leq k} \|D^\alpha u\|_p^p \right)^{\frac{1}{p}} \text{ if } 1 \leq p < \infty$$

$$\|u\|_{k,\infty} = \|u\|_{W^{k,\infty}} = \max_{|\alpha| \leq k} |D^\alpha u|_\infty \text{ if } p = \infty$$

$$W^{s,p}(\Omega) := \left\{ u \in W^{k,p}(\Omega) : \frac{|D^\alpha u(x) - D^\alpha u(y)|}{\|x - y\|^{\sigma + \frac{n}{p}}} \in L^p(\Omega \times \Omega), \forall \alpha, \ |\alpha| = k \right\}$$

$BC^m(J; \mathbb{X}) = \{ f \in C^m(J; \mathbb{X}) : f^{(k)} \in BC(J; \mathbb{X}), \ k = 0, 1, ..., m \}$

$$\|f\|_{BC^m(J;\mathbb{X})} := \sum_{k=0}^m \|f^{(k)}\|_\infty, \quad \forall f \in BC^m(J; \mathbb{X})$$

$$C^{0,\alpha}(J; \mathbb{X}) = \left\{ f \in BC(J; \mathbb{X}) : [f]_{C^{0,\alpha}(J,\mathbb{X})} = \sup_{t,s \in J, s < t} \frac{\|f(t) - f(s)\|}{(t - s)^\alpha} < \infty \right\}$$

$\|f\|_{C^{0,\alpha}(J;\mathbb{X})} = \|f\|_\infty + [f]_{C^{0,\alpha}(J;\mathbb{X})}$

$C^{k,\alpha}(J; \mathbb{X}) = \{ f \in BC^k(J; \mathbb{X}) : f^{(k)} \in C^{0,\alpha}(J; \mathbb{X}) \}$

$\|f\|_{C^{k,\alpha}(J;\mathbb{X})} = \|f\|_{BC^k(J;\mathbb{X})} + [f^{(k)}]_{C^{0,\alpha}(J;\mathbb{X})}$

$$\text{Lip}(J; \mathbb{X}) = \left\{ f \in BC(J; \mathbb{X}) : [f]_{\text{Lip}(J;\mathbb{X})} = \sup_{t,s \in J, s < t} \frac{\|f(t) - f(s)\|}{(t - s)} < \infty \right\}$$

$\|f\|_{\tilde{L}ip(J;\mathbb{X})} = \|f\|_\infty + [f]_{\text{Lip}(J;\mathbb{X})}$

$$C_b^{0,\alpha}(\overline{\Omega}) = \left\{ f : \overline{\Omega} \mapsto \mathbb{C} \text{ bounded continuous, } [f]_{C_b^\alpha(\overline{\Omega})} = \sup_{x \neq y \in \overline{\Omega}} \frac{|f(x) - f(y)|}{\|x - y\|^\alpha} < \infty \right\}$$

$\|f\|_{C_b^{0,\alpha}(\overline{\Omega})} = \|f\|_\infty + [f]_{C_b^{0,\alpha}(\overline{\Omega})}$

$C_b^{k,\alpha}(\overline{\Omega})$—Consists of all functions $f : \overline{\Omega} \mapsto \mathbb{C}$ which are k-times continuously differentiable functions with bounded partial derivatives such that $D^\beta f \in C_b^{0,\alpha}(\overline{\Omega})$

$$\|u\|_{C_b^{k,\alpha}(\overline{\Omega})} = \sum_{|\beta| \leq k} \|D^\beta u\|_\infty + \sum_{|\beta| = k} [D^\beta]_{C_b^{0,\alpha}(\overline{\Omega})}$$

$(\mathbb{X}^*, \|\cdot\|_*)$—Denotes the dual of the normed vector space $(\mathbb{X}, \|\cdot\|)$

$B(\mathbb{X}, \mathbb{Y})$—Denotes the collection of all bounded linear operators from $\mathbb{X}$ to $\mathbb{Y}$

$\|\cdot\|$—Denotes the operator norm defined by $\displaystyle\sup_{x \in \mathbb{X}\setminus\{0\}} \frac{\|Ax\|}{\|x\|}$

$\mathscr{O}(\mathbb{X})$—Denotes the class of invertible linear operator defined $\mathbb{X}$

$N(A) = \mathrm{Ker}\,(A) = \left\{x \in \mathbb{X} : Ax = 0\right\}$—Kernel of A

$R(A) = \left\{Ax : x \in \mathbb{X}\right\}$—Range of A

$D(A)$—Denotes the domain of A

$\mathscr{G}(A) = \left\{(x, Ax) : x \in D(A)\right\}$—Graph of A

$\mathscr{F}(\mathbb{X}, \mathbb{Y})$—Denotes the class of finite rank operators from $\mathbb{X}$ to $\mathbb{Y}$

$\mathscr{K}(\mathbb{X}, \mathbb{Y})$—Denotes the class of compact linear operators

$\|x\|_{D(A)} = \|x\| + \|Ax\|$ for all $x \in D(A)$—Graph norm

$[D(A)] = (D(A), \|\cdot\|_{D(A)})$

$\rho(A) = \left\{\lambda \in \mathbb{C} : (\lambda I - A)^{-1} \in B(\mathbb{X})\right\}$—Resolvent set of A

$\sigma = \mathbb{C} \setminus \rho(A)$—Spectrum of A

$\Phi(\mathbb{X})$—Denotes the collection of Fredholm operators

$\sigma_{ess}(A) = \left\{\lambda \in \mathbb{C} : \lambda I - A \text{ is not a Fredholm operator of index } 0\right\}$—Essential spectrum

$R(\lambda, A) = R_\lambda^A = (\lambda I - \lambda)^{-1}$—Resolvent of A

$S_{\theta,\omega} := \left\{\lambda \in \mathbb{C} : \lambda \neq \omega, \ |\arg(\lambda - \omega)| < \theta\right\}$—Sector

$AP(\mathbb{R}, \mathbb{X})$—Denotes the collection of almost periodic functions

$AP(\mathbb{Z})$—Denotes the collection of almost periodic sequences

$M(f)$—Denotes the mean of f

$AAP(\mathbb{X})$—Denotes the class of asymptotically almost periodic functions

$AAP(\mathbb{Z})$—Denotes the class of asymptotically almost periodic sequences

$bAP(\mathbb{Z} \times \mathbb{Z}, \mathbb{R}^k)$—Denotes the class of bi-almost periodic sequences

$B^2(\mathbb{Z}, \mathbb{R}^N)$—Denotes the class of Besicovitch almost periodic sequences

S_α^β—Denotes an $(\alpha, \alpha)^\beta$-resolvent family

T_α^β—Denotes an $(\alpha, 1)^\beta$-resolvent family

$n := \lceil\beta\rceil$—Denotes the smallest integer greatest than or equal to β

$D_t^\beta u(t) = \dfrac{d^n}{dt^n}\left[\displaystyle\int_0^t g_{n-\beta}(t-s)u(s)ds\right]$, $t > 0$—Riemann–Liouville fractional derivative

$\mathbb{D}_t^\beta u(t) = D_t^{n-\beta} u^{(n)}(t) = \displaystyle\int_0^t g_{n-\beta}(t-s)u^{(n)}(s)ds$—Caputo fractional derivative

$(f * g)(t) := \displaystyle\int_0^t f(t-s)g(s)ds$—Convolution

$g_{\alpha+\beta} = g_\alpha * g_\beta$ for all $\alpha, \beta \geq 0$

$\mathbb{X}_\alpha^A := \left\{x \in \mathbb{X} : \|x\|_\alpha^A := \sup_{r>0} \|r^\alpha (A - \varsigma)R(r, A - \varsigma)x\| < \infty\right\}$

References

1. S. Abbas, M. Benchohra, Partial hyperbolic differential equations with finite delay involving the Caputo fractional derivative. Commun. Math. Anal. **7**, 62–72 (2009)
2. S. Abbas, M. Benchohra, Fractional order Riemann–Liouville integral equations with multiple time delay. Appl. Math. E-Notes **12**, 79–87 (2012)
3. S. Abbas, M. Benchohra, On the set of solutions of fractional order Riemann–Liouville integral inclusions. Demonstratio Math. **46**(2), 271–281 (2013)
4. S. Abbas, M. Benchohra, *Advanced Functional Evolution Equations and Inclusions* (Springer, Cham, 2015)
5. S. Abbas, M. Benchohra, J. Henderson, Global asymptotic stability of solutions of nonlinear quadratic Volterra integral equations of fractional order. Comm. Appl. Nonlinear Anal. **19**(1), 79–89 (2012)
6. S. Abbas, M. Benchohra, G.M. N'Guérékata, *Topics in Fractional Differential Equations*. Developments in Mathematics, vol. 27 (Springer, New York, 2012)
7. S. Abbas, M. Benchohra, T. Diagana, Existence and attractivity results for some fractional-order partial integro-differential equations with delay. Afr. Diaspora J. Math. **15**(2), 87–100 (2009)
8. P. Acquistapace, B. Terreni, A unified approach to abstract linear nonautonomous parabolic equations. Rend. Sem. Mat. Univ. Padova **78**, 47–107 (1987)
9. P. Acquistapace, F. Flandoli, B. Terreni, Initial boundary value problems and optimal control for nonautonomous parabolic systems. SIAM J. Control Optim. **29**, 89–118 (1991)
10. R.A. Adams, *Sobolev Spaces*. Pure and Applied Mathematics, vol. 65 (Academic Press, New York, 1975)
11. R.A. Adams, J.J.F. Fournier, *Sobolev Spaces*. Pure and Applied Mathematics (Amsterdam), vol. 140, 2nd edn. (Elsevier/Academic Press, Amsterdam, 2003)
12. S. Agmon, On the eigenfunctions and on the eigenvalues of general elliptic boundary value problems. Commun. Pure Appl. Math. **15**, 119–147 (1962)
13. L. Amerio, G. Prouse, *Almost-Periodic Functions and Functional Equations* (Van Nostrand Reinhold Co., New York-Toronto, Ont.-Melbourne, 1971)
14. F. Ammar-Khodja, A. Bader, A. Benabdallah, Dynamic stabilization of systems via decoupling techniques. ESAIM Control Optim. Calc. Var. **4**, 577–593 (1999)
15. P.K. Anh, N.H. Du, L.C. Loi, Singular difference equations: an overview. Vietnam J. Math. **35**, 339–372 (2007)
16. J.M. Appell, A.S. Kalitvin, P.P. Zabrejko, *Partial Integral Operators and Integrodifferential Equations*, vol. 230 (Marcel and Dekker, New York, 2000)

17. D. Araya, R. Castro, C. Lizama, Almost automorphic solutions of difference equations. Adv. Differ. Equ., Art. ID 591380, 15 pp. (2009)

18. G. Avalos, I. Lasiecka, Boundary controllability of thermoelastic plates with free boundary conditions. SIAMJ. Control Optim. **38**, 337–383 (1998)

19. D. Baleanu, K. Diethelm, E. Scalas, J.J. Trujillo, *Fractional Calculus Models and Numerical Methods* (World Scientific Publishing, New York, 2012)

20. M. Baroun, S. Boulite, T. Diagana, L. Maniar, Almost periodic solutions to some semilinear non-autonomous thermoelastic plate equations. J. Math. Anal. Appl. **349**(1), 74–84 (2009)

21. E. Bazhlekova, The abstract Cauchy problem for the fractional evolution equation. Fract. Calc. Appl. Anal. **1**, 255–270 (1998)

22. E. Bazhlekova, Subordination principle for fractional evolution equations. Fract. Calc. Appl. Anal. **3**, 213–230 (2000)

23. E. Bazhlekova, Fractional evolution equations in Banach spaces, Ph.D. Thesis, Eindhoven University of Technology, 2001

24. A. Benabdallah, M.G. Naso, Null controllability of a thermoelastic plate. Abstr. Appl. Anal. **7**, 585–599 (2002)

25. S. Benzoni, *Spectre des opérateurs différentiels*. Lecture Notes, 2010

26. A.S. Besicovitch, *Almost Periodic Functions* (Dover Publications, Newburyport, 1954)

27. P. Bezandry, T. Diagana, *Almost Periodic Stochastic Processes* (Springer, New York, 2011)

28. H. Bohr, Zur Theorie der fastperiodischen Funktionen I. Acta Math. **45**, 29–127 (1925)

29. H. Bohr, *Almost Periodic Functions* (Chelsea Publishing Company, New York, 1947)

30. S. Boulite, L. Maniar, G.M. N'Guérékata, Almost automorphic solutions for hyperbolic semilinear evolution equations. Semigroup Forum **71**(2), 231–240 (2005)

31. H. Brézis, *Analyse fonctionnelle: théorie et applications*, 2nd edn. (Masson, 1987)

32. H. Brézis, *Functional Analysis, Sobolev Spaces and Partial Differential Equations* (Springer, New York, 2011)

33. J. Caballero, A.B. Mingarelli, K. Sadarangani, Existence of solutions of an integral equation of Chandrasekhar type in the theory of radiative transfer. Electron. J. Differ. Equ. **2006**(57), 1–11 (2006)

34. S.L. Campbell, Optimal control of discrete linear processes with quadratic cost. Int. J. Syst. Sci. **9**(8), 841–847 (1978)

35. K.M. Case, P.F. Zweifel, *Linear Transport Theory* (Addison-Wesley, Reading, MA, 1967)

36. S. Chandrasekher, *Radiative Transfer* (Dover Publications, New York, 1960)

37. C. Chicone, Y. Latushkin, *Evolution Semigroups in Dynamical Systems and Differential Equations* (American Mathematical Society, Providence, 1999)

38. J.B. Conway, *A Course in Functional Analysis* (Springer, New York, 1985)

39. J.B. Conway, *A Course in Operator Theory*. Graduate Studies in Mathematics, vol. 21 (American Mathematical Society, Providence, 1999)

40. C. Corduneanu, *Almost Periodic Functions*. 2nd edn. (Chelsea, New York, 1989)

41. C. Corduneanu, *Almost Periodic Oscillations and Waves* (Springer, New York, 2010)

42. C. Cuevas, S. Alex, H. Soto, Almost periodic and pseudo-almost periodic solutions to fractional differential and integro-differential equations. Appl. Math. Comput. **218**(5), 1735–1745 (2011)

43. J.M. Cushing, S.M. Henson, Global dynamics of some periodically forced, monotone difference equations. J. Differ. Equ. Appl. **7**, 859–872 (2001)

44. C.M. Dafermos, On the existence and the asymptotic stability of solutions to the equations of linear thermoelasticity. Arch. Ration. Mech. Anal. **29**, 241–271 (1968)

45. T. Diagana, *Pseudo-Almost Periodic Functions in Banach Spaces* (Nova Science Publishers, New York, 2007)

46. T. Diagana, Existence of globally attracting almost automorphic solutions to some nonautonomous higher-order difference equations. Appl. Math. Comput. **219**(12), 6510–6519 (2013)

47. T. Diagana, *Almost Automorphic Type and Almost Periodic Type Functions in Abstract Spaces*, vol. XIV (Springer, New York, 2013), 303 p.

48. T. Diagana, D. Pennequin, Almost periodic solutions for Some semilinear singular difference equations. J. Differ. Equ. Appl. **24**(1), 138–147 (2018)
49. T. Diagana, S. Elaydi, A. Yakubu, Population models in almost periodic environments. J. Differ. Equ. Appl. **13**(4), 239–260 (2007)
50. T. Diagana, H. Henriquez, E. Hernandez, Asymptotically almost periodic solutions to some classes of second-order functional differential equations. Differ. Integr. Equ. **21**(5–6), 575–600 (2008)
51. K. Diethelm, *The Analysis of Fractional Differential Equations* (Springer, Berlin, 2010)
52. N.H. Du, V.H. Linh, N.T.T. Nga, On stability and Bohl exponent of linear singular systems of difference equations with variable coefficients. J. Differ. Equ. Appl. **22**(9), 1350–1377 (2016)
53. Y. Eidelman, V. Milman, A. Tsolomitis, *Functional Analysis: An Introduction*. Graduate Studies in Mathematics, vol. 66 (American Mathematical Society, Providence, 2004)
54. S.N. Elaydi, Nonautonomous difference equations: open problems and conjectures, in *Differences and Differential Equations*. Fields Institute Communications, vol. 42 (American Mathematical Society, Providence, 2004), pp. 423–428
55. K.J. Engel, R. Nagel, *A Short Course on Operator Semigroups* (Graduate Texts in Mathematics, Springer, 2006)
56. K. Fan, Les fonctions asymptotiquement presque-périodiques d'une variable entière et leur application à l'étude de l'itération des transformations continues. Math. Z. **48**, 685–711 (1943)
57. H.O. Fattorini, *Second Order Linear Differential Equations in Banach Spaces*. North-Holland Mathematics Studies, vol. 108 (North-Holland, Amsterdam, 1985)
58. A.M. Fink, *Almost Periodic Differential Equations*. Lecture Notes in Mathematics, vol. 377 (Springer, New York, 1974)
59. J.E. Franke, A.A. Yakubu, Multiple attractors via cusp bifurcation in periodically varying environments. J. Differ. Equ. Appl. **11**(4–5), 365–377 (2002)
60. I. Gohberg, S. Goldberg, M.A. Kaashoek, *Basic Classes of Linear Operators* (Birkhäuser, Bassel-Boston-Berlin, 2003)
61. A. Granas, J. Dugundji, *Fixed Point Theory* (Springer, New York, 2003)
62. M.E. Gurtin, A.C. Pipkin, A general theory of heat conduction with finite wave speed. Arch. Rat. Mech. Anal. **31**, 113–126 (1968)
63. A. Halanay, V. Rasvan, *Stability and Stable Oscillations in Discrete Time Systems*. Advances in Discrete Mathematics and Applications, vol. 2 (Gordon and Breach Science Publication, Amsterdam, 2000)
64. H.R. Henríquez, C. Vásquez, Almost periodic solutions of abstract retarded functional differential equations with unbounded delay. Acta Appl. Math. **57**, 105–132 (1999)
65. D.B. Henry, A.J. Perissinitto, O. Lopes, On the essential spectrum of a semigroup of thermoelasticity. Nonlinear Anal. **21**, 65–75 (1993)
66. E.M. Hernández, Existence of solutions to a second-order partial differential equations with nonlocal conditions. Electron. J. Differ. Equ. **2003**(51), 10 pp. (2003)
67. E. Hernández, H.R. Henríquez, Existence results for second order partial neutral functional differential equation. Dyn. Contin. Discrete Impuls. Syst. Ser. A Math. Anal. **15**(5), 645–670 (2008)
68. E. Hernández, M. McKibben, Some comments on: existence of solutions of abstract nonlinear second-order neutral functional integrodifferential equations. Comput. Math. Appl. **46**(8–9) (2003). Comput. Math. Appl. **50**(5–6), 655–669 (2005)
69. Y. Hino, S. Murakami, T. Naito, *Functional-Differential Equations with Infinite Delay*. Lecture Notes in Mathematics, vol. 1473 (Springer, Berlin, 1991)
70. Y. Hino, T. Naito, N.V. Minh, J.S. Shin, *Almost Periodic Solutions of Differential Equations in Banach Spaces* (CRC Press, Taylor and Francis, Hoboken, 2001)
71. S. Hu, M. Khavani, W. Zhuang, Integral equations arising in the kinetic theory of gases. Appl. Anal. **34**, 261–266 (1989)
72. J.K. Hunter, B. Nachtergaele, *Applied Analysis* (World Scientific Publishing, River Edge, 2001)

73. T. Kato, *Perturbation Theory for Linear Operators*, 2nd edn. (Springer, Berlin, 1976)
74. C.T. Kelly, Approximation of solutions of some quadratic integral equations in transport theory. J. Integr. Equ. **4**, 221–237 (1982)
75. V. Keyantuo, C. Lizama, M. Warma, Existence, regularity and representation of solutions of time fractional diffusion equations. Adv. Differ. Equ. **21**(9–10), 837–886 (2016)
76. M.A. Khamsi, W.A. Kirk, *An Introduction to Metric Spaces and Fixed Point Theory*. Pure and Applied Mathematics (New York) (Wiley-Interscience, New York, 2001)
77. A.A. Kilbas, H.M. Srivastava, J.J. Trujillo, *Theory and Applications of Fractional Differential Equations* (Elsevier Science B.V., Amsterdam, 2006)
78. C. Kuratowski, Sur les espaces complètes. Fund. Math. **15**, 301–309 (1930)
79. G. Lebeau, E. Zuazua, Decay rates for the three-dimensional linear system of thermoelasticity. Arch. Ration. Mech. Anal. **148**, 179–231 (1999)
80. H. Leiva, Z. Sivoli, Existence, stability and smoothness of a bounded solution for nonlinear time-varying thermoelastic plate equations. J. Math. Anal. Appl. **285**, 191–211 (2003)
81. B.M. Levitan, *Almost-Periodic Functions* (G.I.T–T.L., Moscow, 1959). (in Russian)
82. B.M. Levitan, V.V. Zhikov, *Almost Periodic Functions and Differential Equations* (Moscow University Publishing House, 1978). English Translation by Cambridge University Press, 1982
83. J.H. Liu, G.M. N'Guérékata, N.V. Minh, *Topics on Stability and Periodicity in Abstract Differential Equations*. Series on Concrete and Applicable Mathematics, vol. 6 (World Scientific, Singapore, 2008)
84. C. Lizama, J.G. Mesquita, Almost automorphic solutions of non-autonomous difference equations. J. Math. Anal. Appl. **407**(2), 339–349 (2013)
85. J. Locker, *Spectral Theory of Non-self-adjoint Differential Operators*. AMS Mathematical Surveys and Monographs, vol. 73 (American Mathematical Society, Providence, 2000)
86. L. Lorenzi, A. Lunardi, G. Metafune, D. Pallara, *Analytic Semigroups and Reaction Diffusion Problems*. Internet Seminar, 2004–2005
87. A. Lunardi, *Analytic Semigroups and Optimal Regularity in Parabolic Problems*. Progress in Nonlinear Differential Equations and their Applications, vol. 16 (Birkhäuser, Basel, 1995)
88. A. Lunardi, *Linear and Nonlinear Diffusion Problems*. Lecture Notes, 2004
89. L. Maniar, R. Schnaubelt, *Almost Periodicity of Inhomogeneous Parabolic Evolution Equations*. Lecture Notes in Pure and Applied Mathematics, vol. 234 (Dekker, New York, 2003), pp. 299–318
90. C.M. Marle, *Mesures et probabilités* (Hermann, Paris, 1974)
91. K.S. Miller, B. Ross, *An Introduction to the Fractional Calculus and Differential Equations* (Wiley, New York, 1993)
92. M.G. Naso, A. Benabdallah, Thermoelastic plate with thermal interior control, in *Mathematical Models and Methods for Smart Materials (Cortona, 2001)*, Series on Advances in Mathematics for Applied Sciences, vol. 62 (World Science Publisher, River Edge, 2002), pp. 247–250
93. A.W. Naylor, G.R. Sell, *Linear Operator Theory in Engineering and Science*. Applied Mathematical Sciences, vol. 40 (Springer, New York, 1982)
94. G.M. N'Guérékata, *Almost Automorphic Functions and Almost Periodic Functions in Abstract Spaces* (Kluwer Academic/Plenum Publishers, New York/London/Moscow, 2001)
95. G.M. N'Guérékata, *Topics in Almost Automorphy* (Springer, New York, 2005)
96. J.W. Nunziato, On heat conduction in materials with memory. Q. Appl. Math. **29**, 187–204 (1971)
97. J.T. Oden, L.F. Demkowicz, *Applied Functional Analysis*, 2nd edn. (CRC Press, Boca Raton, 2010)
98. B.G. Pachpatte, Volterra integral and integrodifferential equations in two variables. J. Inequ. Pure Appl. Math. **10**(4), 1–21 (2009)
99. A. Pankov, *Bounded and Almost Periodic Solutions of Nonlinear Operator Differential Equations* (Kluwer, Dordrecht, 1990)

100. A. Pazy, *Semigroups of Linear Operators and Applications to Partial Differential Equations*. Applied Mathematical Sciences, vol. 44 (Springer, New York, 1983)
101. I. Podlubny, *Fractional Differential Equations* (Academic Press, San Diego, 1999)
102. W. Rudin, *Functional Analysis* (Springer, New York, 1976)
103. S.G. Samko, A.A. Kilbas, O.I. Marichev, *Fractional Integrals and Derivatives. Theory and Applications* (Gordon and Breach, Yverdon, 1993)
104. R. Schnaubelt, Sufficient conditions for exponential stability and dichotomy of evolution equations. Forum Math. **11**, 543–566 (1999)
105. R. Schnaubelt, Asymptotic behavior of parabolic nonautonomous evolution equations, in *Functional Analytic Methods for Evolution Equations*, ed. by M. Iannelli, R. Nagel, S. Piazzera. Lecture Notes in Mathematics, vol. 1855 (Springer, Berlin, 2004), pp. 401–472
106. C.C. Travis, G.F. Webb, Compactness, regularity, and uniform continuity properties of strongly continuous cosine families. Houst. J. Math. **3**(4), 555–567 (1977)
107. C.C. Travis, G.F. Webb, Cosine families and abstract nonlinear second order differential equations. Acta Math. Acad. Sci. Hungaricae **32**, 76–96 (1978)
108. A.N. Vityuk, A.V. Golushkov, Existence of solutions of systems of partial differential equations of fractional order. Nonlinear Oscil. **7**, 318–325 (2004)
109. W. von Wahl, *Gebrochene potenzen eines elliptischen operators und parabolische differentialgleichungen in Räumen höldersteliger Funktionen*, Nachr. Akad. Wiss. Göttingen, Math.-Phys. Klasse **11**, 231–258 (1972)
110. R.N. Wang, D.H. Chen, T.J. Xiao, Abstract fractional Cauchy problems with almost sectorial operators. J. Differ. Equ. **252**(1), 202–235 (2012)
111. J. Weidmann, *Linear Operators in Hilbert Spaces* (Springer, New York, 1980)
112. K. Yosida, *Functional Analysis* (Springer, New York, 1965)
113. S. Zaidman, *Almost-Periodic Functions in Abstract Spaces*. Research Notes in Mathematics, vol. 126 (Pitman/Advanced Publishing Program, Boston, 1985)
114. C. Zhang, *Almost Periodic Type Functions and Ergocity* (Kluwer Academic Publishers, Dordrecht, 2003)

Index

A

Abstract phase space, 151
Acquistapace-Terreni conditions, 52
Adherent point, 4
Adjoint operator, 35
Almost periodic, 58, 61, 63, 140
Almost periodic sequence, 67
Almost sectorial operators, 125
Analytic semi-group, 48
Anti-linear, 25
$AP(\mathbb{X})$, 58
Arzelà-Ascoli theorem, 8
Asymptotically almost periodic, 65, 66, 139, 151
Autonomous difference equation, 75

B

Banach fixed point theorem, 8
Banach space, 12
$BC(J; \mathbb{X})$, 19
$BC^m(J; \mathbb{X})$, 20
β-times integrated semi-group, 128
Bi-almost periodicity, 76
Bijective, 31
Bounded operator, 29
Brouwer fixed point theorem, 23
$B(\mathbb{X}, \mathbb{Y})$, 29

C

C_0-semi-group, 46
$C_b^{0,\alpha}(\overline{\Omega})$, 21
$C_b^{k,\alpha}(\overline{\Omega})$, 21
$C^\alpha(J; \mathbb{X})$, 20

Caputo fractional derivative, 126
Cauchy problem, 125
Cauchy-Schwarz inequality, 25
Cauchy sequence, 4, 5
Circular spectral mapping theorem, 51
$C^{k,\alpha}(J; \mathbb{X})$, 20
Classical solution, 114, 119, 125
Closable, 33
Closed, 33
Closed ball, 3
Closed set, 3
$C^m(J; \mathbb{X})$, 20
Compact metric space, 7
Compact resolvent, 40
Complete metric space, 5
Continuous function, 6
Continuous linear operator, 29
Continuous spectrum, 37
Convergence, 4
Convolution, 15, 126
Cosets, 12
Cosine family, 151
Countable, 11

D

Dense, 11
Densely defined, 35
Derivative operator, 34
Diameter, 4
Discrete exponential dichotomy, 77
Discrete logistic equation, 10
Discrete metric, 2
Domain, 33
Dual space, 22

© Springer Nature Switzerland AG 2018

T. Diagana, *Semilinear Evolution Equations and Their Applications*,
https://doi.org/10.1007/978-3-030-00449-1

E

Embedding theorems, 17
Energy relaxation function, 152
ε-period, 58
ε-translation, 58
Equi-continuous, 8
Equilibrium points, 10
Essential spectrum, 38
Evolution family, 51
Exponential dichotomy, 53
Exponential growth bound, 51
Exponentially bounded, 51
Exponentially stable, 51

F

Fading memory, 139, 151
Fourier coefficients, 63
Fourier exponents, 64
Fourier series, 62, 64
Fractional Cauchy problem, 125
Fractional derivative, 128
Fredholm operator, 38

G

Generalized Pythagorean theorem, 26
Gradient, 17, 151
Graph, 33
Graph norm, 33

H

Heat conduction, 139, 151
Heat flux, 151
Hilbert space, 24
Hille-Yosida Theorem, 48
History, 151
Hölder space, 20, 21
Hyperbolic semi-group, 50

I

Infinitesimal generator, 46
Injective, 31
Inner product, 25
Interior point, 3
Internal energy, 151
Invertible, 31

K

Kernel, 34

L

Laplace operator, 34
Laplace transform, 126
Left shift operator, 30
Leray-Schauder alternative, 24
Linear operator, 29
$Lip(J; \mathbb{X})$, 20
Lipschitz space, 20
Logistic equation, 10
$L^p(\Omega)$, 14
$L^p(\Omega, \mu)$, 13

M

Mean value, 62
Measure, 13
Measurable sets, 13
Measure space, 14
Metric, 2
Metric space, 2
Mild solution, 114, 117, 120, 125
Mittag-Leffler special function, 127

N

Non-autonomous difference equation, 75

O

Open ball, 3
Open covering, 7
Open set, 3
Orthogonal, 25
Orthogonal projection, 26

P

Point spectrum, 37
Projection theorem, 26
Pythagorean theorem, 25

Q

Quotient normed vector space, 13
Quotient space, 12

R

Range, 34
Rapidly decreasing function, 46
Real interpolation space, 53
Relatively compact, 7
Relatively dense, 58

Residual spectrum, 37
Resolvent, 39
$(\alpha, 1)^\beta$-resolvent families, 128
$(\alpha, \alpha)^\beta$-resolvent families, 125, 128
Resolvent set, 36
$\rho(A)$, 36
Riemann-Liouville fractional derivative, 126
Right shift operator, 30

S
Sadovsky fixed-point theorem, 23
Scalar product, 25
Schauder fixed point theorem, 23
Schwartz space, 46
Sectorial operator, 40, 53
Semilinear evolution equation, 140
Separable, 11
Sequential continuity, 6
Sesquilinear form, 24
$\sigma(A)$, 36
σ-algebra, 13
Sobolev space, 15, 16
Spectrum, 36
Sphere, 3
$\mathscr{S}(\mathbb{R}^n)$, 46
Stress relaxation function, 152
Strict contraction, 9
Strict solution, 113, 114, 119
Strongly continuous semi-group, 45
Strongly stable, 50

Strong solution, 114
Sup-norm, 19
Surjective, 31

T
Temperature, 151
Thermoelastic plate systems, 139
Transition matrix, 77
Triangle inequality, 2
Trigonometric polynomial, 58

U
Unbounded delay, 139
Unbounded linear operator, 33
Uncountable, 11
Uniformly bounded, 8
Uniformly continuous, 6
Uniformly convex, 11
Uniformly exponentially stable, 50
Uniformly stable, 50

V
Variation of temperature, 140
Vertical deflection, 140

W
$W^{k,p}(\Omega)$, 16

Printed in the United States
By Bookmasters